W9-AGA-164

Japanese-Trained Armies In Southeast Asia

By the same author:

Chandora Bosu to Nihon (in Japanese), Tokyo: Hara Shobo (1968)

Jungle Alliance, Japan and the Indian National Army, Singapore: Asia/Pacific Press (1971)

Okuma Shigenobu, Statesman of Meiji Japan, Canberra: Australian National University Press (1973)

Japan's Greater East Asia Co-Prosperity Sphere in World War II: Selected Readings and Documents, Oxford University Press (1975) edited

Women in Changing Japan, Boulder, Colorado: Praeger/Westview Press (1976) co-edited with Joy Paulson and Elizabeth Powers

Japanese-Trained Armies in Southeast Asia

Independence and Volunteer Forces in World War II

Joyce C. Lebra

Columbia University Press
New York 1977

Published in 1977 in Hong Kong, Singapore, and Malaysia by Heinemann Educational Books (Asia) Ltd., and in the United States of America by Columbia University Press

Printed in Hong Kong

Library of Congress Cataloging in Publication Data

Lebra, Joyce C
Japanese-trained armies in Southeast Asia

Bibliography
Includes index
1. Asia, Southeastern – Armed Forces – History. 2. Military education – Asia, Southeastern – History. 3. Asia, Southeastern – History – Japanese occupation
I. Title
UA830.L4 1977 355.3'5 75-16116
ISBN 0-231-03995-6

Contents

Preface

I spent 1965-66 in India studying the Indian National Army. The INA was a revolutionary army which, with the aid of the Japanese, fought the British in Southeast Asia during World War II. The INA experience, though a part of the history of the Indian Independence Movement, is little known in the U.S. I also spent four summers in Japan studying Japanese policy towards wartime India and the INA. The results of this research were published first in Japanese: *Chandora Bosu to Nihon* (Hara Shobo, 1968), then in English: *Jungle Alliance, Japan and the Indian National Army* (Singapore: Asia/Pacific Press, 1971).

In the course of this research it became apparent that Japanese military training bore fruit in more than the INA. In fact, Japan trained liberation or volunteer forces in most nations of wartime Southeast Asia. I therefore embarked on research on these armies during a two-year period, 1970-72, which I spent in Japan, Southeast Asia, and Australia. At the same time I studied the more general framework within which these armies were fostered, the Greater East Asia Co-Prosperity Sphere.

I also became aware that a number of Western scholars, using a behavioral approach, had devised typologies for the role of the military in developing nations. Some scholars, for example, have pointed to the "transference of military skills to civilian administration in new nations".[1] Others have constructed typologies for role expansion of the military.[2] Still others have written generally of the relationship between

[1] Janowitz, Morris, *The Military in the Political Development of New Nations.*

[2] See, for example, Lissak, Mosche, "Modernization and Role Expansion of the Military in Developing Countries", in *Comparative Studies in Society and History,* v.9, 1966-67, no.3, pp.233-56.

military and political power in new states.[3]

These studies demonstrate that the role of the military as a modernizing and westernizing influence has become a fruitful area of research. Emphasis has been on typologies and model-building. When specific examples are cited, they are often chosen from armies in the new nations of Africa. When Southeast Asian armies are considered, there is often no mention of Japanese training, but rather focus on earlier British or Dutch conditioning. The history of the British Indian Army is well recorded, at least from the British viewpoint, over a three-century span, because of the predilection of British officers for writing memoirs and recollections. The Burma Rifles and *KNIL* in Indonesia have achieved a similar immortality through the eyes of British and Dutch officers. The influences selected in model-building scholarly writing tend to be from these examples or, if in new states, often from Africa.[4]

There has been a curious neglect of empirical studies of Japanese-trained armies in Southeast Asia. We can explain this partly, perhaps, in view of the scarcity of relevant linguistic skills among Western scholars. It is nevertheless noteworthy, given the penchant of Western social scientists to study colonial regimes and even military history, that few have turned to the military impact of Japan on the rest of Asia. A small number of studies have been made of the effects of Japanese military administration on individual nations of Southeast Asia. Few if any Western scholars, however, have dealt with the phenomenon of Japanese-trained independence armies, which in some cases have formed the nuclei of the officer corps of the armies and also of political élites in post-war Southeast Asian nations.

In Korea, in the Empire proper, Japanese wartime military recruitment of cadres of young men directly into the Imperial Army is similarly a neglected aspect of historical

[3] For example, Huntington, S., *Changing Patterns of Military Politics;* Gutteridge, W., *Military Institutions and Power in New States;* and Johnson, John, ed., *The Role of the Military in Underdeveloped Countries.*

[4] For recent scholarly studies, for example, see Lucien Pye's article, "The Army in Burmese Politics", in Johnson, ed., *The Role of the Military in Underdeveloped Countries;* Gutteridge, W., *Military Institutions and Power in New States;* and Von der Mehden, Fred, *Politics of the Developing Nations.*

writing on that nation. Professor Hahn-been Lee, for example, describing the training of the officer corps of the Korean Armed Forces, overlooks their Japanese training and emphasizes instead the Military English Language School, "established in late 1945 by the United States Military Government with a view to preparing the cadre for a future army of an independent country".[5]

I hope in this volume to remedy some of this scholarly oversight and to provide data and stimulus for future research in this fascinating and critical area.

[5] Lee, Hahn-been, *Korea: Time, Change and Administration*, p.145. An exception to this oversight is Se-jin Kim, who in *The Politics of Military Revolution in Korea*, considers the role of Japanese-trained officers in the Korean Army. In Korea as elsewhere in Asia the militarization of a Japanese-trained cadre meant not only the genesis of a later Korean Army but also of post-war political leadership. The post-liberation power élite of Korea includes significant numbers of men of this Japanese experience. Among those who attended the Military Academy in Japan is President Park. Ex-Premier Chung Il-kwon was trained at the Japanese Military Academy in Manchukuo, as were others. Equally significant constellations of Japanese-trained military officers fill the top ranks of Korea's military establishment, though they may not be eager to admit to their Japanese experience.

Acknowledgements

For support during research from 1970 through 1972 in Japan, Southeast Asia and Australia, I am indebted to the National Endowment for the Humanities, the American Philosophical Society, and the Australian National University. All provided fellowships for research on the Greater East Asia Co-Prosperity Sphere. During 1965-66 I received a Fulbright Grant to support my research in India on the Indian National Army. During the summers intervening between 1966 and 1970 I received travel support for research in Japan from the American Philosophical Society and the University of Colorado.

The individuals who have given generously of their time, encouragement and advice are too numerous to mention here. A few of those in Japan, the U.S. and Australia without whose gracious counsel and encouragement this study could not have been completed are gratefully acknowledged here: Mr. Horie Yoshitaka, Professor Oka Yoshitake, Professor Hayashi Shigeru, Dr. Tsunoda Jun, Professor Masuda Atō, Professor Ohno Torhu, Professor Ichikawa Kenjirō, Professor Ogiwara Hiroaki, Colonel Imaoka Yutaka, Colonel Fujita Yutaka, General Fujiwara Iwaichi, Mrs. Shiraishi Aiko (née Kurasawa), Mr. Kurokawa Nobuo, Mr. Nakamura Mitsuo, Mr. Nishihara Masashi, Mr. Nakano Keiji, Miss Hirano Midori, Admiral Tomioka Sadatoshi, General Miyamoto Shizuo, Colonel Ochiai Shigeyuki, Mr. Hirano Jirō, Mr. Fujino Yukio, Mr. Kawashima Takenobu, Mr. Takahashi Hachirō, Mr. Maruyama Shizuo, Professor Homma Shirō, Professor Itagaki Yoichi, Mr. Yamashita Masao, Mr. Kawadji Susumu, Dr. Anthony Reid, Miss Enid Bishop, Professor Yoji Akashi, Professor Lawrence Beer, and Professor Roger Paget.

1

Japanese Initiative – Southeast Asian Response

Suharto, Ne Win and Park all have something in common. Of course all three are Asian leaders who head military bureaucracies respectively in Indonesia, Burma and Korea. They have another bond, much less well known but at least equally significant. These three generals were all trained during World War II in Japanese-fostered military units and schools. These men were part of a far-flung series of independence and volunteer armies organized and trained by the Japanese during the Pacific War, primarily in Southeast Asia. This training has affected a whole generation of leadership in Southeast Asia.

Two independence armies, the Indian National Army and the Burma Independence Army, were especially significant in size and in wartime and post-war impact. In Indonesia, the Army of Defenders of the Homeland, or *Peta (Sukarela Tentara Pembela Tanah Air)* has had a powerful influence on political leadership and on the Indonesian Army. In other areas volunteer armies were organized and trained which have had varying degrees of influence on post-war leadership in Sumatra, Malaysia, Indochina and the Philippines.

Another feature of the Japanese military imprint was the organization and training, though only minimally military, of numerous paramilitary groups, for example in Java, armed mainly with bamboo spears. While the military training was minimal the numbers of young men involved were considerable (see Appendix B). Thousands of Southeast Asians were also recruited as *heiho,* auxiliary troops used in many subsidiary functions. Even more significant than the independence armies themselves in terms of the quality of military training imparted was the selection of talented Southeast Asians from the independence and volunteer armies for officer training at the Military Academy in Japan. The imprinting of the Japanese military model on the military bureaucracies of Southeast Asia is a highly significant phenomenon which this manuscript attempts to delineate.

These Japanese-trained armies have become agents of revolution and modernization in several post-war nations, stimulating new leadership groups which guided their nations toward independence. This result was partly fortuitous from the Japanese standpoint and even from the Southeast Asian vantage point. The ramifications of the creation of these armies went far beyond their original goals and have become a kind of time bomb in Southeast Asia.

Today the officer corps in many Southeast Asian nations is the major source of political power and administrative ability. It is therefore relevant to look for the causes and origins of this phenomenon of military political power in Asia. The genesis of the rising political significance of the military in Southeast Asia derives in part from the colonial period. But of at least equal, though less recognized, significance is World War II and what happened during Japanese occupation.

Armies, whether in Asia, Latin America, Africa or the Middle East, have often operated as forces for modernization and westernization. Armies have also acted as transmitters of nationalism, diffusing nationalism and anti-colonialism among the populace at critical junctures. Armies may be agents in the early creation of large-scale organizations, in promoting a rational outlook and technological training, in diffusing education and literacy, in encouraging efficiency, in affording a channel for upward social mobility, and in fostering an awareness of the potentialities of mass political action.[1] These Japanese-trained armies thus assume a significance beyond a traditional military role.

Historians have often viewed civil and military authority as mutually incompatible. The wartime and post-war experience in Southeast Asia, however, does not support this

[1] See discussion by Lucien Pye, "Armies in the Process of Political Modernization," pp.69-81, in John J. Johnson, ed., *The Role of the Military in Underdeveloped Countries.*

analysis. In the *Peta* and the Burma Independence Army and its successors, the Burma Defense Army, BDA, and Burma National Army, BNA, a military cadre was almost simultaneously trained and politicized, then fed into the immediate post-war political élite.

Japanese military occupation interjected a stimulating influence into nascent nationalist revolutions in Southeast Asia. Educated nationalists were given Japanese military training and discipline to support and assist the Japanese military effort by defending themselves. Since the enemies of Japan and Asian nationalists were temporarily identical it was possible to forge limited alliances against Western colonial regimes. In at least three cases an élite officer corps was recruited and trained, partly in Southeast Asia and in part at the Military Academy in Japan. Military action against Western colonial authority was the occasion for catalytic Japanese military training.

A negative aspect of Japan's wartime impact was that Japan failed to control or delimit Southeast Asian nationalism and anti-colonialism, as the French, Dutch and British did also and as the U.S. has failed to do since. These forces were directed against Japanese military administration in later phases of the war, as they had been against Western colonial régimes. With Japanese defeat ended any further successful colonial claim to rule in Southeast Asia. These were lessons which the U.S. could have learned but did not from these colonial experiences in Southeast Asia.

Japan had basically two sets of motives in giving military training to a whole generation of nationalists in Asia between 1940 and 1945. In the immediate pre-war months a high priority goal in Tokyo was to encourage disaffection from Britain in Malaya and Burma, among overseas Indians as well as among local nationalists. This motive led to intelligence operations as part of a propaganda war directed primarily against Britain. Japanese intelligence officers were assigned to Malaya and Burma, where they were to contact nationalist leaders and foster anti-British sentiment. Ultimately these intelligence officers determined that one of the best ways to achieve their missions was to organize and train military units which would engage in joint military operations against the

Allies while promoting anti-colonial sentiment. To this propaganda goal were added the strategic imperatives of cutting off India and Australia by sea and in Burma of intercepting the flow of supplies being sent over the Burma Road to the beleaguered Chungking Government. The consequence of this assessment was the organization and training of volunteer armies whose primary goal was to shore up coastal defenses and to supplement Japanese units in anticipation of Allied attack. Thus, volunteer units were trained in Sumatra, Malaya, the Philippines and Indochina, beginning in some cases in late 1943 and continuing until the end of the war. The case of the *Peta* in Java was intermediary between the two categories of armies trained by the Japanese. Both political and strategic considerations contributed to the creation of this volunteer army in Java.

Although these motives influenced the Japanese decisions to create these armies, it was more the lack of a clear, coherent policy in Tokyo toward Southeast Asia that enabled forces within the Army to take the initiative in giving groups of nationalists military training and experience which equipped them later to lead independent armies in post-war Asia. Japan's policy vacuum concerning Southeast Asia was a result of the Army's traditional obsession with the military threat from the Soviet Union, an obsession which led Japan to focus attention on Manchuria, north China and Korea at the expense of other parts of Asia. Lack of policy was also due to Japan's spectacularly rapid success in occupying Southeast Asia. Even as war enveloped all of Southeast Asia the Imperial Army had extreme difficulty operating as a coordinated military machine there.

This is not to say that Japan had no interests in Southeast Asia. There were compelling economic imperatives that caused military men, particularly within the Navy, to look southward for access to rich oil reserves and sources of tin and rubber which Japan's industrial complex desperately needed. By contrast with Japan's colonial and military experience in Manchuria, Korea and Taiwan, nevertheless, Japan was caught off guard, faced with having to devise occupation policies in Southeast Asia with almost no preparation or prior experience. Economic pressures for access to raw materials led the cabinet by late July 1940 to proclaim the goals and existence of the Greater East Asia Co-Prosperity Sphere, embracing all of Southeast Asia to the western boundary of Burma.

While Imperial General Headquarters did have some military, strategic and political goals *vis à vis* India and Burma, Tokyo did not seriously contemplate including India within the bounds of the Greater East Asia Co-Prosperity Sphere. Tokyo Headquarters procrastinated for two years over the strategy of offensive operations inside the Indian border. The Indo-Burma border delineated the Western defense perimeter of the Greater East Asia Co-Prosperity Sphere, and strategic plans projected the occupation of Burma, at least in part. While Tokyo harbored political and propaganda objectives for India, where Burma and Java were concerned, there were clearer military imperatives.

Even Burma was a late-comer to map maneuvers in Tokyo, far overshadowed along with the rest of Southeast Asia by the Army's traditional apprehension over Russia. The image of the Russian threat was orthodox within Army General Staff Headquarters and helped precipitate Japan's involvement in the Russo-Japanese War and annexation of Korea. Japan's expansion into Manchuria and north China also reflected this Army ideology and primary concern.

In the decade preceding 1940 for Japanese foreign policy Asia was China and Manchukuo. Southeast Asia was still beyond the pale, and Pan-Asian ideological constructs of the thirties focussed on the "core area" – Japan, China and Manchukuo. Rationalizations for such constructs as the East Asia Federation and the East Asia Cooperative Community applied only to the core area, invoking geographic and regional ties, affinity of race, script and culture, and economic symbiosis. Western imperialist control, the Versailles system, the Washington-London naval ratios, and and Chinese nationalism were all viewed as obstacles to Pan-Asian co-existence. Only after August 1940 were these arguments and concerns extended to Southeast Asia.

When the Pacific War erupted and Japan overran the countries of Southeast Asia in dramatic succession, Japan was faced with having to devise policies for military occupation.

Japan's pre-war contacts and experience in Southeast Asia were basically private and commercial. Japan had in Southeast Asia nothing like the colonial experience in Taiwan and Korea or the military intelligence network of the early decades of the century in China. In these latter areas thousands of Japanese had staffed civil and military bureaucracies, economic and development enterprises, and fought in ground combat in the twentieth century. While on the one hand Japan lacked similar experience and expertise for dealing with Southeast Asia, on the other hand there were fewer preconceptions or biases, more room to maneuver, and the possibility of devising *ad hoc* pragmatic solutions for local problems.

In the creation of the Indian National Army and the Burma Independence Army propaganda goals and local initiative were uppermost. Both these armies came into being even before the outbreak of war in the Pacific. Their creation was the product of intelligence missions headed by two staff officers from Tokyo Headquarters, Major Fujiwara Iwaichi, heading the *Fujiwara Kikan* (Fujiwara Agency) in Thailand and Malaya, and Colonel Suzuki Keiji, leading the *Minami Kikan* (Southern Agency), in Burma. The effectiveness of these intelligence missions depended on the local initiative exercised by the *kikan* head and his staff. Both these men had been trained in intelligence and were products of the Imperial Army policy of giving wide latitude to field grade officers to carry out important assignments, a policy without counterpart in other armies elsewhere. The scope of these intelligence missions was not delineated in Tokyo but rather depended on the requirements of local day-to-day operations. Fujiwara and Suzuki, if not actually exceeding their orders from Tokyo, were certainly using their own imagination and initiative in acting as midwives of independence armies.

Fujiwara and Suzuki were among about a thousand intelligence agents trained for the most part in the Army Intelligence School, the *Nakano Gakkō*, and sent to Southeast Asia in the guise of diplomats, journalists and businessmen in the months before the war. Intelligence training, always a part of the modern Imperial Army, was formalized with the establishment of the *Nakano Gakkō* in

1940 at the old Signal Corps Center. Most graduates were assigned to *tokumu kikan,* "special duty agencies", all of which were to a degree out of control by Tokyo of necessity. *Kikan* staff members contacted local nationalists, using Japanese military attachés as intermediaries. Colonel Tamura Hiroshi in the Japanese Embassy in Bangkok, for example, had a wide network of local contacts which proved useful to both Fujiwara and Suzuki.

These intelligence agencies and officers were aided locally in their assignments by Japanese residents in Southeast Asia and in Tokyo by various research groups which in some cases acted as lobbies in promoting certain policies and interests in Asia. Best known of these research institutes was the Research Bureau and its subsidiary, the East Asia Research Institute, of the South Manchuria Railway Company. There were many others, such as the South Seas Economic Research Institute, the Burma Research Society, and the Shōwa Research Society, which sought to influence policy toward Southeast Asia prior to and during the war. Some of these "policy companies" were semi-official, receiving covert funding from the Army, Navy or Foreign Ministry. In some cases they also commanded private support and funding. Many books, pamphlets, and position papers were published under the aegis of these groups, who had the expertise of both professional military men and civilian authorities from universities and journalism. This network of research organizations operated within a shadowy but nevertheless significant area in terms of political and military decision-making.

Between this complex set of Japanese forces and Southeast Asian nationalist movements a few Japanese intelligence officers worked to recruit and train these armies. There were some similarities among the Japanese officers who inspired the armies. Fujiwara and Suzuki were later paralleled by Yanagawa Munenari, who helped to organize and train the *Peta.* Each of these men was sent out on an intelligence mission to establish liaison with nationalist movements. Each was imbued with a romantic and idealistic self-image of his own and Japan's role in fostering Indian, Burmese, or Javanese independence. All three were struck by parallels between their own missions and that of the legendary Lawrence of Arabia.[2] In this romantic idealism – uncharacteristic of the general war-time stereotype of the Japanese officer in Southeast Asia – Fujiwara, Suzuki and Yanagawa were remarkably similar. Through the imaginative negotiations and prompting of these men three revolutionary armies were spawned. Fujiwara and Suzuki established a degree of harmony with nationalist leaders and credibility for Japanese motives that were never equalled by their successors. Some Indian, Burman and Indonesian participants still recall with nostalgia this early period of Japanese encouragement and have revisited Japan since the war.[3]

Fujiwara, and even more Suzuki, was impatient with bureaucratic channels through which he had to thread his way in bringing his project to life. Both men displayed great initiative and imagination, and demonstrated a genuine commitment to the ideal of independence, a goal and commitment not shared in equal degree by their superiors in the chain of command. Both the *F Kikan* and the *Minami Kikan* were under the command of Southern Army Commander Marshal Terauchi and his staff, and below that under the 15th Army Staff in Burma. The frustrations Fujiwara and Suzuki expressed in working through the military bureaucracy shortened their careers as heads of intelligence agencies in Southeast Asia. Tokyo's reaction to Fujiwara was not as abrupt as toward Suzuki, for Suzuki paid less attention to the chain of command than Fujiwara. Suzuki's insistence on setting up an independent Burmese government and immediately liberating all of Burma caused eyebrows to be raised in Tokyo, in Burma and in Saigon.[4]

[2] Interviews with Fujiwara Iwaichi, summer 1965, Tokyo; and with the son of Suzuki Keiji, 21 November 1970, Hamamatsu City. See also Yanagawa Munenari, *Rikugun chōhōin Yanagawa chūi* (Lt. Yanagawa, an Army intelligence officer).

[3] Interviews with Mohan Singh in Delhi, 1965-66. General Ne Win, one of the original thirty young Burmans given Japanese military training, visited Japan in the spring of 1970 and saw old wartime comrades. Earlier, he invited Col. Suzuki to visit Burma. Sukarno and Suharto have also visited Japan in recent years.

[4] Interviews with Takahashi Hachiro, 5 Oct. 1970; Sugii Mitsuru, 17 Nov. 1970; and Maj.-Gen. Nasu Yoshio, 8 Oct. 1970, all in Tokyo.

Suzuki also faced within the *Minami Kikan* a more complex situation than confronted Fujiwara in the *F Kikan.* Fujiwara's staff consisted initially of a small handful of men imbued with a close sense of esprit and camaraderie. The *Minami Kikan,* on the other hand, reflected in microcosm the rivalry between the Army and Navy, for both were involved in the creation of the *Minami Kikan.* There were in addition serious personal rivalries among the members. When Tokyo decided that an intelligence agency had exceeded its bounds or outlived its usefulness, it was either abolished, as in the case of the *Minami Kikan,* or reorganized in somewhat altered form, as with the *F Kikan* and its successor, the *Iwakuro Kikan.* The *tokumu kikan* were a big gamble for Tokyo, with potentially great gains and losses at stake.

Why did Tokyo completely abolish the *Minami Kikan* but allow the *F Kikan* to continue and in fact to expand its activities? Both policies reflected a reimposition of control by Tokyo over free-wheeling field agencies. The difference arose from the fact that the *F Kikan* was after all dealing with Indians outside their own country, with a tenuous base of support. With Burmese nationalists on their own soil there was too large an unknown factor for Tokyo to reckon with. Had the *F Kikan* been operating within India and under Japanese occupation it might have suffered the fate of the *Minami Kikan.* There was also a difference in Tokyo's attitude towards Fujiwara and Suzuki. Though both men had risen from within General Staff Headquarters, Suzuki had already earned a reputation for causing trouble in Shanghai and was at one stage confined to his home for a month as punishment. Yet Suzuki had the qualities of an effective intelligence operative. Fujiwara had a clearer appreciation of military protocol and did not ignore specific orders from Tokyo as Suzuki did in attempting to carry out his difficult and intricate plans. Still, the limits of Fujiwara's project were not specifically defined, and this ambiguity and Fujiwara's idealistic passion for Indian independence in fact led to his replacement by a hard-headed political officer, Colonel Iwakuro Hideo.

Abolition of the *Minami Kikan* and transferring Suzuki back to Japan were also evidences of Tokyo's reassessment of the situation in 1942. Another evidence of greater caution was the abolition of the BIA and its recreation as the smaller Burma Defense Army, a carefully selected and more professionally trained cadre. 1942 saw a dual Tokyo policy toward independence armies of military retrenchment coupled with expansion of political and administrative responsibilities.

The INA and BIA had already been in operation several months when at the end of June, 1942, there was some indication of interest at the policy level in Tokyo. Continental Directive 1196 issued by Imperial General Headquarters to the Headquarters of the Southern Army in Singapore specified, "In important parts of the Southern Theater, in order to facilitate the execution of new duties we will train necessary armed groups."[5] This order, with the precedent of prior actions of Fujiwara and Suzuki, established the fundamental stance of IGHQ on the formation and training of volunteer armies of local inhabitants in Southeast Asia.

Following this order it was over a year before Southern Army Staff Headquarters took any action on Continental Directive 1196. One reason for the inaction was that no deadline was set in the order. Another significant reason was that the Southern Army Headquarters felt no sense of urgency in June or July of 1942. Still another factor in the inaction was the reduction in the size of Staff Headquarters of the Southern Army after occupation of Southeast Asia in early 1942. The number of staff sections was reduced from four to two, and later again rose to three. This contrasted with the size of the staff of the Kwantung Army in China, which at the time included five sections. The delay was therefore partly a matter of lack of personnel to assign to a project whose urgency was not immediately apparent.[6]

It was not until a year later, in September 1943, that Southern Army Staff Headquarters issued a follow-up order to its own subordinate armies in Southeast Asia. The order

[5] Interview with Col. Imaoka, 28 Feb. 1972. See Imaoka Yutaka, *Nansei hōmen rikugun sakusen shi* (Campaign history of the Army in the Southeast Asia sector), unpublished, War History Library.

[6] Interview with Col. Imaoka, 28 Feb. 1972.

called for formation of armies of native inhabitants in each part of Southeast Asia.[7] Action was finally taken at the initiative of Lieutenant-General Inada Masazumi, assigned from Tokyo to be Deputy Chief of Staff of the Southern Army in February 1943. In June Inada made an inspection tour of Southeast Asia, including Java, Sumatra and Borneo in his itinerary. Inada became apprehensive about Japan's deficient defensive strength, realizing that troop reinforcements could not be spared from higher priority areas in the Pacific where decisive campaigns were being fought. Inada made specific recommendations to Commander General Harada of the 16th Army in Java and to General Tanabe of the 25th Army in Sumatra for the training of corps of local inhabitants to supplement Japanese troop strength in these areas. Inada discussed his plan with Generals Tōjō and Satō Kenryō of Tokyo Staff Headquarters who were also touring Southeast Asia in the summer of 1943, and these generals gave Inada's plan their imprimatur.[8] Execution of the details of the project was left to the staffs of the various armies under the command of the Southern Army, and was promoted by the local commanding officers.

Another stage in the evolution of Japanese Southeast Asia policy was reached in 1943, the year of Japanese-proclaimed "independence" for Burma and the Philippines. It was a year when the pace of Japan's military thrust was definitely faltering. The slowing of the momentum was accompanied by some re-evaluation in Tokyo of political warfare goals. If continued Japanese military success was not a constant which could be guaranteed, then the propaganda effort would have to be stepped up. The result of this assessment was the promise in January of independence for Burma and the Philippines within the year, and the calling of the Greater East Asia Conference in November in Tokyo. Burma delebrated "independence" accordingly on 1 August and the

[7] *Ibid.*

[8] George Kanahele alleges that Tōjō refused funds or arms for the project. See *The Japanese Occupation of Indonesia,* p.120. Inada, however, in an interview with the author, denied that Tōjō refused funds and said that he, Inada, then made arrangements in Tokyo to implement the plan: interview 24 Nov. 1970.

Philippines on 14 October. It was, however, a qualified independence, reminiscent of the Japanese-sponsored state of Manchukuo. Filipino leadership reflected awareness of this in its lack of enthusiasm for Japanese-style independence, though Jose Laurel and other leaders had their own motives for working with Japanese authorities. The chief limitations related to Japan's military and economic imperatives. And while Japan sent Sawada Renzō as Ambassador to the new Burma, this did not satisfy the aspirations of nationalistic Thakins for complete independence.

Japan's qualified encouragement of Southeast Asian independence in the 1943 phase was well illustrated by the Greater East Asia Conference in Tokyo in the first week of November. Some fifty leading nationalists from all Southeast Asia were invited to attend. It was a major propaganda offensive and marked the apogee of the good will effort of Japan's whole Southeast Asia program. The calling of the Conference also commemorated the first anniversary of the creation of the Greater East Asia Ministry. Subhas Chandra Bose, Ba Maw, Wang Ch'ing-wei and Laurel were among the delegates who addressed the assemblage. Bose, Ba Maw and others in their speeches expressed gratitude for Japan's assistance in expelling white imperialism from Asia.

Tōjō spoke to the delegates, celebrating the new age of harmony and co-prosperity in an Asia which had expunged the parasitic and oppressive influence of Western colonialism, marking the dawn of "Asia for Asians". A major product of the Conference was a resolution drafted by the Greater East Asia Ministry. This resolution epitomized both the propagandistic nature of the Conference and the height of good will within the Greater East Asia Co-Prosperity Sphere.

Tokyo's revived interest in independence for Southeast Asians prompted the arrival in Asia of the charismatic revolutionary, Subhas Chandra Bose, who did not hesitate to press demands on Tōjō and Sugiyama in Tokyo. Bose's position as head of the Free India Provisional Government and his value in reviving and unifying the Indian independence movement and INA led Tokyo to take some measures toward supporting the Indian government-in-exile. Japan recognized the new government, sent a diplomatic envoy, and transferred to the government the captured Andaman and Nicobar Islands, symbolic as a political prison under the British. All these measures, however, were more symbolic than substantive and were designed chiefly for their propaganda value.

1943 saw too the beginning of the execution of Inada's plan of forming volunteer armies, the largest and most significant being the *Peta.* While "independence" was being heralded in other parts of Southeast Asia, 16th Army Commander General Harada in Java announced on 3 October 1943 the creation of *Peta.* The Japanese Army had, as General Inada had noted, only a meager 10,000 troops in the 16th Army. Lt. Yanagawa and his intelligence staff went into action to recruit and train the *Peta* officer corps.

The timing of the formation of the INA and BIA on the one hand and the volunteer armies, the *giyūgun* on the other, reflects some fundamental differences in conception and purpose from the Japanese standpoint. First, the INA and BIA were fundamentally part of a propaganda war in which political considerations outweighed military in the Japanese rationale. It was in Japan's interest to encourage Indian and Burmese aspirations for independence from England, even though the process of the formulation of Japan's independence policy for India and Burma was often tortuous and attended by disagreement within and between all levels of the Japanese military command.

In the case of the *Peta* in Java the imperatives for propaganda warfare were not clear as with the BIA and INA. The reason was that independence for Java or Indonesia as a whole was not a clear policy goal of Japan during the war. Demands for independence by Indonesian nationalists were therefore hedged with Japanese equivocation, attempting to sustain Indonesian hopes without ever granting them in substance. Segments of the Japanese military opposed any hint of independence. The Navy in occupation of the Indonesian Outer Islands and some in IGHQ and in the 25th Army in Sumatra looked toward permanent possession of part or all of Indonesia. Other sections of the Japanese military structure, even though they may have favored some concessions to Indonesian aspirations, could, therefore, not achieve consensus on this issue. Still, the 16th Army took cognizance of the political and propaganda implications of its administration of Java. This was reflected in many ways, including the fact that the Japanese officers who trained and advised *Peta* were, like those who trained the INA and BIA, graduates of the *Nakano Gakkō*. Japan in Java also sponsored numerous mass organizations for political mobilization of support for Japanese occupation policies.

This political rationale did not apply in Malaya, Sumatra, Borneo or other areas in Southeast Asia, where smaller volunteer armies were formed. In these parts of Southeast Asia there was no Japanese intent to foster or encourage nationalism or independence aspirations among the inhabitants. Nor were nationalist movements as well developed as in Burma, Java, or among overseas Indians. The volunteer units created in those areas were not, therefore, part of political warfare in the sense that the INA and BIA were. The Malayan and Sumatran units had more limited objectives, and the officers who trained them were not *Nakano Gakkō* alumni. These armies had specifically military objectives, functioning in roles supplementary to Japanese units. The *Peta* provides a kind of intermediary example in the spectrum of Japanese goals in the indigenous armies, between the INA and BIA on the one hand, and the volunteer units in Malaya, Sumatra and elsewhere on the other.

Japanese political goals in encouraging independence movements and armies were affected by the serious military reverses of late 1943 and 1944. After Guadalcanal it was apparent that Japan no longer had a real hope of regenerating a second offensive. This was a "campaign first" period when the Japanese Army had one overriding concern: defense of areas under occupation and stopping the military progress of the Allies in the Pacific. In late 1943 the Army sought a rousing victory somewhere which would give a desperately needed boost to morale at home. Tokyo was expecting an Allied counter-offensive in Burma. IGHQ also wanted to use Bose because of his power with Indian opinion. The INA meanwhile aspired to cross the border into India to ignite revolution there, starting in Bengal. These imperatives finally dictated approval for the Imphal offensive, the only major campaign which tested the limits of cooperation between the INA and Japan. The campaign, finally mounted in 1944, was a fiasco (see Chapter Two) and dashed Indian hopes of any further help from Japan. The BNA played no major role in this offensive, though much of the fighting took place in Burma. The reason was that the 15th Army did not have the faith in the BNA as a battle-ready force that it had in the INA.

During 1944, as Japan's military weakness was increasingly exposed in Southeast Asia, culminating finally in the desperate last-ditch measures of the campaign of Leyte Gulf, a corresponding shift in attitudes of nationalist leaders toward Japan was discernible. The luster of Japan's early military successes was dimmed by her inability to defend the areas she had occupied initially, as well as by Allied gains. As occupying power, Japan had come to take on some of the repressive features of her Western colonial predecessors. By the end of 1944 doubts about the role of Japan as liberator of Asia and dissatisfaction with the rigors of military occupation had created unrest of explosive proportions in some areas. Already in August 1944, there were several incidents of violence between Burman and Japanese and Sumatran and Japanese soldiers. 1945 was marked by open revolt against Japan by Japanese-trained units in Burma and Java. But in these nationalist-inspired revolts against their Japanese tutors the armies demonstrated that they had become genuine independence armies with revolutionary potential.

An essential ingredient of these armies, and one over which the Japanese had little control, was Southeast Asian nationalism, which sought to use the Pacific War as a chance for liberation from Western colonial rule. Interaction of Japanese and endemic forces produced nascent armies where, in many cases, there were no significant colonial antecedents. The Indian National Army with its long colonial tradition in the British Indian Army in this respect was exceptional and contrasted with the BIA in Burma and the *Peta* in Java, where their colonial predecessors trained negligible numbers of Indonesians and Burmans. The real difference between these Javan and Burmese independence armies and their colonial predecessors, however, was not size but rather function. Colonial forces in pre-war Burma and Indonesia performed primarily internal police functions. Japanese-trained units were also recruited from different segments of the populace, engendering new élites with new skills and expertise.

Fujiwara in creating the Indian National Army worked with the Indian Independence League, which had an extensive network of agents in Malaya and Burma. During the actual organization of the INA he recruited from among Indian prisoners of war from the British Indian Army, who surrendered in the jungles of Malaya and by the thousands at Singapore. Suzuki recruited the original Burmese "thirty comrades" from among young student activists of the Thakin Party. Among these original thirty were Thakin Aung San, post-war premier assassinated with his cabinet, and Thakin Shu Maung, better known as Ne Win. The thirty were selected and sent to Hainan and Taiwan for training during 1940 by Suzuki and members of the *Minami Kikan,* most of whom were graduates of the *Nakano Gakkō.*

Among recruits from both the Indian Independence League and the Thakin Party, nationalist aspirations for independence were critical in the decisions to cooperate with the Japanese and to join military units which would equip them to fight for independence. Even with the Indian prisoners of war, latent anti-British sentiment had been fed by discriminatory British policies within the British Indian Army, which treated British and Indian officers differently in terms of promotion, pay, and access to officers' clubs. Captain Mohan Singh, commander of the original Indian National Army, confessed to Fujiwara to having these feelings. Many surrendered Indians at Singapore, however, were deterred from cooperating with the Japanese and joining the INA by long family traditions of loyalty to the British. Some of these men later overcame their reluctance after the arrival in Asia of Subhas Chandra Bose, who had escaped from British India to Nazi Germany.

Those who joined the volunteer armies were also nationalists first, and though Japan did not promote nationalism and independence movements in these areas, the

Malays, Sumatrans and Filipinos who "collaborated" did not believe they were betraying their national interests. Were these men simply puppets and traitors, as they have often been depicted in Western literature?[9] Collaboration is no black and white issue, though it has often been presented simplistically. Motivation and pressures to join were complex, and as with most men who wield political power, venality and opportunism were mingled with genuine patriotic devotion. Among nearly all – those who headed governments-in-exile in Washington, London or India and those who threaded their precarious way through the Japanese military bureaucracy in Rangoon, Djakarta, Singapore and Manila – nationalism was preponderant.

The stigma adhering to those who collaborated was in part engendered by returning Western colonial powers. The ambivalence of the position of those who opted to remain in their Japanese-occupied homelands was generally acknowledged with empathy both by those nationalists who left and those who remained. There was no universal stigma of collaboration in the eyes of most Southeast Asians. Many who held office under Japanese occupation have on the contrary been hailed as heroes by their compatriots. Subhas Chandra Bose, Aung San, Ne Win, Sukarno and Suharto have been acclaimed as real patriots and revolutionaries against Western rule. In Djakarta and Manila on the eve of Japanese occupation there were agreements among the leadership, Sukarno and Hatta, Laurel and Vargas agreeing to remain while others resisted from exile. The caretakers resisted within the framework of cooperation, refusing Japanese demands and requests as they were able, cooperating when necessary.

The stigma of collaboration was also in part a function of the strength of local independence movements and of the duration of the struggle of the Western colonial powers against Japanese invasion. In the Philippines, where the U.S.

[9] For exceptions to this simplistic characterization see discussion of this problem in Boyle, John Hunter, *China and Japan at War, 1937-1945, The Politics of Collaboration;* Lebra, Joyce C., *Jungle Alliance, Japan and the Indian National Army;* and Steinberg, David Joel, *Philippine Collaboration in World War II.*

had already promised independence and American defense was protracted, some stigma persisted. In Java, on the other hand, where the Dutch had given no promises and defense had lasted only eight days, nationalist leaders who worked with the Japanese have been praised for persevering in the independence struggle. Similarly in Burma, where the British had not only not promised independence but had imprisoned and exiled young Thakin Party leaders, there was not much acrimony after the war. "The British left us to fend for ourselves,"[10] one Burman remarked.

Recruits in the independence and volunteer armies saw in cooperation with the Japanese the only real hope for independence. When these hopes were dashed by repressive Japanese policies the armies turned to fight their Japanese tutors, in Java, in Burma, and in Sumatra. Officers and men trained in these Japanese-fostered armies and units, far from being vilified for collaboration, have risen in the vanguard of political and military leadership after the war.

[10] Regret has been expressed by some Englishmen as well, not only for British departure but for the manner of that withdrawal. W.G. Burchett, for example, in *Trek Back From Burma,* writes, "I don't know why we didn't offer these people in Malaya and Burma some good reason to fight for themselves", p.22.

2

The Indian National Army

Japan at the outbreak of war in the Pacific faced critical problems for which she was ill prepared. Southeast Asia was a newcomer to map maneuvers in the upper echelons of Army General Staff Headquarters in Tokyo, where attention was traditionally focussed rather on North China and Soviet Russia. Even less was Japan prepared to formulate policy toward India, for there were no India experts in Headquarters, and only a few civilian authorities in Japan. There was nothing that could be called an "India lobby" in Japan. Even the Foreign Ministry was represented diplomatically in British India only by consular staff in the major cities, and the Foreign Minister in the cabinet was losing ground to the service ministers and Chiefs of Staff in the struggles in Liaison Conferences where important wartime decisions were forged. India was a peripheral concern for Japan in 1941.

To counteract the lack of information about South and Southeast Asia in Tokyo, the Second Bureau of Imperial General Headquarters adopted the expedient of sending young field-grade officers to every nation in Southeast Asia in a series of far-flung intelligence missions. The information fed back from these agencies became the basis for policy decisions made within Imperial General Headquarters. Other ingredients in the wartime decision-making process were the preconceptions of officers in the First Bureau, the powerful Operations Bureau, which overrode the Second Bureau in any confrontation.

One of the most important independence armies created by Japan in Southeast Asia was the Indian National Army, spawned within the framework of an inchoate policy toward India. What were Japanese policy calculations regarding India? First, Japan never seriously contemplated encompassing India within the bounds of the Greater East Asia Co-Prosperity Sphere, though some grosser versions of the "Greater Sphere", or sphere of influence, did include India. The design was for a sphere which would sweep through Southeast Asia to the Indo-Burma border, and even in the case of Burma the original operational plans did not include all of Burma. The Greater East Asia Co-Prosperity Sphere would expunge Western colonialism from Southeast Asia and replace it with an economically self-sufficient entity under Japanese control, supplying critical oil, tin and rubber for Japan's industrial establishment.

India was of military significance to Japan as the western defense perimeter of the Sphere. India was also the origin of supplies to Chinese and American forces in Chungking. The Indian National Army was from the Japanese viewpoint primarily a propaganda means of fostering anti-British sentiment among soldiers and civilians in Southeast Asia and within India.[1] As Japan sought to expel British influence from India and the rest of Southeast Asia, so also Indian nationalists aspired to free India. Out of this common enemy arose the limited cooperation between Japan and India that focussed on the Indian National Army.

From several sources Tokyo learned of the existence in Berlin of an Indian revolutionary exile, Subhas Chandra Bose. In April 1941, for example, Consul General Okazaki in Calcutta described in a secret communiqué to Foreign Minister Matsuoka the activities of a radical party in Bengal, the Forward Bloc. Okazaki suggested establishing contact with this party in India and with its leader, Subhas Chandra Bose, in Berlin. Okazaki felt that the movement might burgeon into a popular revolt, and he advised tangible Japanese aid, including weapons, to the party.[2]

[1] This is also the conclusion of K.K. Ghosh in *The Indian National Army, Second Front of the Independence Movement.* For greater detail on the INA and its relationship to Japan see Lebra, *Jungle Alliance, Japan and the Indian National Army.* The author, because of prior publication *of Jungle Alliance,* has decided to place less emphasis here on the INA than on the Japanese-trained armies.

[2] Gaimushō Kiroku, *Dai Toa senso kankei ikken, Indo mondai* (Matters relating to the Greater East Asia war, India problem), Secret communiqué from Okazaki to Matsuoka, nos. 11975-11978, 11979, 30-31 April 1941.

From Ambassador General Oshima Hiroshi in Berlin also came communiqués regarding Bose and his hopes to go to East Asia. Though Bose had helped organize an Indian Legion of ex-POWs from North Africa in Germany, by late 1941-early 1942 Bose had begun to despair of German aid for the Indian independence struggle and to suggest to Oshima that Japan cooperate with Indian independence leaders against Britain. These two sources of information on Bose in the Foreign Ministry in Tokyo were supplemented at the end of 1941 by Army intelligence reports from Southeast Asia. But meanwhile the Foreign Ministry made no decision on Bose.

Japan's minimal concern with India was reflected in several decisions and declarations in Tokyo in late 1941 and early 1942. On 15 November 1941, a Liaison Conference decision, the "Plan for Acceleration of the End of the War with America and Britain" called among other things for: "1) separation of Australia and India from Britain, and 2) stimulation of the Indian independence movement."[3] Problem for Japan with regard to the second objective was the Gaimushō assessment that the Indian National Congress was anti-Japanese. Another problem was the belief that, even if the Indian independence movement should succeed, it would be difficult for Indian revolutionaries to establish a stable, orderly state. Nor would it be possible for Japan to control a nation of four hundred million in addition to her other commitments in Southeast Asia.[4] But it did lie within the realm of feasibility for Japan to launch a vast propaganda effort to encourage Indian disaffection from Britain. This was in fact the policy Japan adopted toward India.

On the other hand, Indo-Japanese cooperation in Southeast Asia drew on a cultural nexus sanctioned by tradition. For Japan, India was the home of Hinduism, of the historical Buddha, and of the modern saint of nationalism, Gandhi. As the source of inspiration for centuries of artistic, literary and philosophical expression, India attracted devout Buddhists and literary figures. Okakura Tenshin and Rabindranath

[3] Ike, Nobutaka, *Japan's Decision for War, Records of the 1941 Policy Conferences*, p.247.

[4] Secret document signed Ott (German Ambassador in Tokyo), 7 January 1942, Tokyo, IMTFE, Exhibit.

Tagore both celebrated Asian spirituality as a heritage which was distinctively Asian.

For Indians, as for many other Asians, Japan was a source of encouragement for aspirations for freedom from colonial rule. Japan's spectacular victory over the sprawling giant, Russia, in 1905, infused hope into several generations of Asian nationalists before the Pacific War. Japanese patriotic societies and individuals had for over two decades protected revolutionary exiles from many parts of Asia from the authorities of colonial regimes. Bose and other Indian revolutionaries during World War II, therefore, responded positively to Japan's call for "Asia for Asians", for ridding Asia of Western imperialist control.

Tōjō declared in the Diet early in 1942, "Without the liberation of India there can be no real mutual prosperity in Greater East Asia", and further, in April, "It has been decided to strike a decisive blow against British power and military establishment in India."[5] Tōjō mentioned India in Diet speeches on January 17, February 12, February 14, March 11-12, and April 4. Repeatedly he called on Indians to take advantage of the war to rise against British power and establish an India for Indians. These were, however, general policy statements rather than directives to the Operations Bureau in IGHQ. The actual problem of a strategic operational plan for crossing the Indo-Burma border was much more complex and consumed hundreds of hours of time of staff officers in Tokyo and in Burma in 1942, 1943, and 1944.

At several points, it was conceivable that a Japanese invasion of India might have had some success had it been planned and executed. The optimum time was in the spring and summer of 1942, following Japanese successes in Malaya and Burma, when Japanese air, sea and land power could not readily have been checked by the British. Japan passed up the opportunity. Again, in the fall of 1942, a thrust into India was contemplated, "Operation 21" as it was called. But in

[5] Tōjō speech on military activities in India: Imperial Conference Decision 4 April 1942, in *Dai Hon'ei seifu renraku kaigi kettei gijiroku* (Records of the Imperial General Headquarters-Government Liaison Conference Decisions), War History Library.

1942-43, there were too many obstacles to the idea – inadequate supply lines, British deterrent strength in the Akyab sector through early 1943, a shortage of trained Indian troops for a joint campaign, not to mention events in the Pacific. The plan was consequently postponed to 1944, when the chances of success were even less, as some cautioned when the fateful decision was actually made.

The Imphal campaign as finally envisaged had the limited objective of "securing strategic areas near Imphal and in Northeast India for the defense of Burma".[6]

Apart from the disastrous Imphal campaign, Japan's wartime policy toward India was largely concerned with the Indian National Army and government-in-exile, the Free India Provisional Government. On 1 October 1941, Fujiwara Iwaichi, a thirty-three year-old major with the 8th section, Second Bureau, IGHQ, was sent to Bangkok on a highly secret intelligence mission. Chief of Staff General Sugiyama, who had himself spent two years in India, instructed Fujiwara in his formidable assignment: to contact the Indian independence movement, the overseas Chinese, and the Malay sultans with a view to encouraging friendship and cooperation with Japan. He was given a staff of five commissioned officers and two Hindi-speaking interpreters.

Fujiwara was dumbfounded at the scope of his mission, for which he felt ill prepared. But assigning missions of such magnitude to field-grade officers and then allowing them involvement with China and Korea, out of fear of Russia, the command structure and tradition of the Japanese Army. This policy had no counterpart in the British, American, or Third Reich armies.

Fujiwara also felt ill prepared because he knew little about India or Southeast Asia. Here again his lack of preparation reflected the situation of the Japanese Army generally. There were no India specialists in IGHQ and only a few travel guides to India in the IGHQ library. The Army's traditional involvement with China and Korea, out of fear of Russia, meant that Japan was ill prepared to deal with either India or

[6] Instructions from Imperial General Headquarters to General Kawabe Shōzō in Burma 7 January 1944, in the Historical Section, Defense Ministry, Government of India, New Delhi.

Southeast Asia in 1940.

Still, Fujiwara was a conscientious, capable, and dedicated officer who threw himself totally into his mission. At first, Fujiwara reported to the Japanese military attaché in Bangkok, and later to the 25th Army and Southern Army Headquarters.

Fujiwara began work contacting groups of Indians in Bangkok. Inspiration for the organization of the INA grew out of talks between Fujiwara and two Sikhs: Pritam Singh, a religious leader and teacher, who headed the network of the Indian Independence League in Malaya and Thailand, and Mohan Singh, one of the first Indian officers captured by the Japanese in central Malaya. Fujiwara and Mohan Singh took an immediate liking to each other, and Fujiwara convinced Mohan Singh that he would be treated as a friend and ally, not a prisoner of war. In his imaginative appeal to join forces with Japan, Fujiwara pointed to historic and cultural ties between India and Japan and urged Mohan Singh that the opportunity to fight for Indian freedom had arrived. The three men worked out a formula for contacting all Indians in the British Indian Army in Southeast Asia to persuade them to volunteer for the INA. The INA was born at the end of December 1941, and Fujiwara watched as Mohan Singh was transformed into a passionate revolutionary.[7]

Mohan Singh immediately began to train units to work beside those groups already operating under Pritam Singh's direction in the jungles of Malaya and Thailand. Mohan Singh organized all officers into a single class, adopted common kitchens, slogans and songs. These steps helped to create revolutionary *élan* and to transcend, if briefly, traditional communal antagonisms and rivalries. The insignia of the INA uniform was a springing tiger. The size of the INA prior to Singapore was about 2,500. It was during this period that Fujiwara established with the Indians the initial credibility

[7] These negotiations which attended the birth of the INA are described in unpublished manuscripts in the possession of Mohan Singh, and by Fujiwara Iwaichi in *F Kikanchō no shuki* (Memo of the chief of the *F Kikan*) Jieitai, Tokyo, 1959 and in Fujiwara, *F Kikan*, as well as in Toye, Hugh, *Subhash Chandra Bose, the Springing Tiger;* and in Lebra, *Jungle Alliance.* See Mohan Singh's now published accounts.

for Japan's sincerity in cooperating with the Indian independence movement. Fujiwara and Mohan Singh together created this nucleus of a revolutionary army, British by conditioning and experience, partly Japanese in impetus, and more Indian in character than the British Indian Army. The interaction between Fujiwara and Mohan Singh was, in the words of one *F Kikan* member, "better than watching a *kabuki* drama".

By 8 January, when Tokyo sent an officer from the 8th Section to investigate the progress of Fujiwara's project, the latter already had his own convictions about Japan's India policy. He suggested: 1) Japanese encouragement of the Indian independence movement to cut India adrift from England; 2) clarification of Japan's basic policy toward India and the independence movement; 3) a unified policy toward India; 4) expanding the work of the *F Kikan* to all areas of Asia; 5) a world-wide scope for Japan's India policy, including inviting Bose to East Asia from Berlin; 6) Japanese aid to both IIL and INA; 7) personal proof to Indians in occupied areas of the ideals of the New Order in East Asia, and 8) reorganization and expansion of the *F Kikan* to accomplish these goals.[8]

As a result of these conversations and proposals, Fujiwara was soon visited by two powerful generals from Tokyo, Tominaga and Tanaka. While he was called on to defend his optimistic prognostications regarding the INA and Mohan Singh, he was gratified to see that Tokyo was taking more note of India. He spoke of the creation of a huge army of 100,000 men and expressed the hope that Subhas Bose could be brought from Berlin to lead it.

With Japan's smashing success at Singapore in a single week in February, Fujiwara accepted from the British the surrender of some 45,000 Indian troops used to defend that British bastion. About half this number volunteered for the INA when Fujiwara and Mohan Singh addressed the POWs at a mammoth rally in Farrer Park. The other half, beset by conflicts of loyalty and inability to abandon long family traditions in the British Indian Army, refrained. Some of

[8] Fujiwara, *F Kikanchō no shuki,* pp.134-5; and *F Kikan,* pp.183-6.

these volunteered later for various reasons, especially after the arrival of Subhas Chandra Bose.

In March, representatives of the IIL and INA were called to a conference to Tokyo. Fujiwara was distressed that one of two planes carrying Indian delegates, including Pritam Singh, crashed *en route* to Tokyo and all aboard were lost. At the same time, Fujiwara spent three days discussing with IGHQ the most effective methods of dealing with the Indian independence movement. He felt he had made some headway in modifying the Machiavellian assumptions from which Tokyo was operating. A schism nevertheless remained between IGHQ and Fujiwara, who, while he was genuinely committed to the goal of Indian independence, could not convince IGHQ to share his convictions.

One problem which confronted Fujiwara in Tokyo was the rift between the Indian communities in Tokyo and Southeast Asia, each of which felt they were the true revolutionaries in the independence struggle. This split later caused serious problems for the whole project of Japanese liaison with the independence movement. Head of the Indian delegation in Tokyo was Rash Behari Bose, long-time political exile in Tokyo with a Japanese wife and a son in the Japanese Army. Rash Behari was never completely accepted as an Indian patriot by the Southeast Asian Indian community. The problem was not resolved until Subhas Chandra Bose arrived in Southeast Asia during the summer of 1943 and united the whole Indian independence movement in Asia.

One consequence of Fujiwara's proposals was that he himself was replaced by Colonel Iwakuro Hideo and the *Iwakuro Kikan*. Iwakuro was a politically powerful officer whose career included a role in the Japanese-American negotiations in Washington in 1941 and assisting in founding the Army Intelligence School, the *Rikugun Nakano Gakkō*. Tōjō, reluctant to have such a politically formidable officer in Tokyo, had sent him to Southeast Asia earlier, and Iwakuro now became Fujiwara's successor. The *Iwakuro Kikan* included some two hundred and fifty members, with two Diet members, a far cry from the handful of dedicated men with whom the *F Kikan* began. A few months later, the size had risen to five hundred, with several sub-sections. This expansion reflected Tokyo's increasing stress on the significance of the propaganda war. Iwakuro's personal approach to the Indian independence movement, reflecting his technical expertise, was less idealistic and romantic than Fujiwara's.

Iwakuro was immediately plagued by a vexing problem which Fujiwara had worried about but been unable to resolve – antagonism between Indians in Tokyo and those in Southeast Asia. This mutual suspicion grew until it created a crisis, personified in a struggle between Mohan Singh and Rash Behari Bose. Precipitating cause of the crisis was a series of Indian demands, in the "Bangkok resolution", product of a conference in Bangkok. Indian representatives of the IIL and INA, meeting at Bangkok, drew up a thirty-four point resolution. They wanted a Japanese point-by-point reply. Tokyo was unable to give all the assurances requested in the resolution; furthermore Iwakuro himself advised an attitude of general acceptance. This response was unacceptable to Mohan Singh and his Council of Action. Iwakuro, working closely with Rash Behari, was unable to effect a compromise with Mohan Singh. In Mohan Singh's eyes Bose was nothing more than a Japanese puppet. The result of this confrontation was Mohan Singh's arrest and detention for the duration of the war. More important, the INA virtually ceased to function.

Both Fujiwara and Iwakuro operated on only very general instructions from Tokyo. This gave them room to maneuver but also not as much support as they required from Tokyo at times. Most serious was the ambiguity Iwakuro felt about his role and uncertainty in Tokyo itself about how far Japan should go in support of Indian independence. Fujiwara urged full and sincere support of the independence movement, but IGHQ had many reservations, some of them based on practical problems of higher priority demands for material support for Pacific campaigns. Iwakuro was working from an IGHQ attitude of grudging and limited support, but the problem still remained of determining the limits. In general, Iwakuro read the mood in Tokyo well. The one point which was patent, about which Tokyo could not equivocate, was that the India project was part of a secret war in which the weapons of intelligence and espionage played a key role. Intelligence and secret diplomacy were an old story to Iwakuro, areas where he had proven his versatile talents. But the IIL, INA, and especially Mohan Singh, continually plagued Iwakuro with specific demands, pushing the limits of Japan's willingness and capacity to commit herself. This fundamental problem of defining Japan's policy limits persisted under Iwakuro and ultimately led to the dissolution of the first INA.

With the arrest of Mohan Singh, the INA collapsed. No one was willing to accept leadership, given the relationship with the *Iwakuro Kikan.* In Feburary 1943, some halting efforts were made to revive the INA. A Military Bureau of six Indian officers held the organization together until the arrival of Subhas Chandra Bose in the summer of 1943.

The decision to send Bose from Berlin to Tokyo was reached only after protracted discussions between Tokyo and Berlin. Tokyo at first showed little interest in attempting to bring a Bengali revolutionary to Tokyo and Southeast Asia. There was already another Indian revolutionary in Tokyo named Bose, but his presence had caused the Japanese and Iwakuro, in particular, some problems. It was an IGHQ communiqué to the Japanese military attaché in Berlin that finally prompted negotiations, and even then the joint German-Japanese decision to send Bose to Tokyo was slow in coming.

Subhas Chandra Bose was a revolutionary of palpably greater stature than Rash Behari, and his value was apparent immediately to those Japanese who met him. His leadership of the overseas Indian community was undisputed. Even within India, his political stature had enabled him to be elected President of the Indian National Congress twice, despite the opposition of Gandhi. His revolutionary predilections extended back to his school-days and his rejection of an appointment in the Indian Civil Service. Bose's premise was that the only way to destroy a government was to refuse to cooperate with it. Convinced that the only way to expel Britain from India was through the use of force, he broke with Gandhi and Nehru. He was put under house arrest by the British authorities.

Bose arrived in Tokyo in June 1943, after a secret trip and rendezvous between German and Japanese submarines. His charismatic personality had an immediate impact on Chief of Staff Sugiyama and Foreign Minister Shigemitsu, and even more significantly on Tōjō. After pleading with Sugiyama and Shigemitsu, Bose was able to get what he most wanted in Tokyo: an interview with Tōjō, followed by a second interview.

Once Sugiyama and Shigemitsu met Bose they were convinced that here was a revolutionary of a different ilk from Rash Behari. When Bose met Tōjō, the Prime Minister was likewise impressed by Bose's charisma and dedication. There would have to be some reassessment of Japan's policy toward India and the INA. It was no longer advisable to ignore the political value of the INA and the propaganda value of professing support for Indian independence.

The ensuing shifts in policy derived in part from Bose's personal impact. There was also an altered military situation with the planning and execution of the Imphal offensive. The *Kikan* was reorganized as the *Hikari Kikan*[9] briefly under Colonel Yamamoto Bin, who had been military attaché in Berlin, then under Lieutenant-General Isoda Saburō. Isoda outranked his predecessors, reflecting the increased emphasis on military aspects of liaison in 1944. Isoda was also a benign, gentle-mannered general, whose appointment was designed to placate Bose's insistent demands for action in India. Bose was dissatisfied at having to deal with the *Hikari Kikan;* he would have preferred to deal directly with the Japanese Army and government.

Bose as commander of the INA (but without personal military rank) made some innovations in the INA. When he announced the formation of a women's unit, the Rani of Jhansi Regiment, Japanese military authorities in annoyance refused Bose a camp site. Women had no place in the Japanese martial tradition or in any army, in the Japanese view. Bose persisted until he won his point, but the women's regiment remained a headache for the Japanese. The Rani of Jhansi Regiment was an INA innovation without British or Japanese precedent and helped create a national character

[9] *Hikari* means "light", in this case implying light from the East.

and revolutionary image for the INA.

Bose on arrival in Southeast Asia also formed a Free India Provisional Government on 21 October and became its head. Two days later the FIPG was recognized by Japan, Italy, Germany and several other nations. Bose in November asked Tōjō to cede the Andaman and Nicobar Islands to the FIPG so that it would have territory and be qualified as a government in international law. Tōjō at first demurred, on grounds that the islands were strategically necessary and the Navy would never consent. When Tōjō later did announce the transfer of the islands, it was more as a symbolic gesture. Though Bose sent a commissioner to the islands, the Navy never relinquished control.

Bose also asked Tōjō to send a diplomatic envoy. IGHQ demurred and never completely agreed. Tōjō again arranged a compromise gesture. An experienced diplomat Hachiya Teruya was sent, but when he arrived in Southeast Asia Bose refused to meet him, as he had no credentials and was therefore not a regularly accredited diplomat.

The Japanese devoted little effort to actual military training of the INA, since they were already experienced soldiers. The one area where INA training was felt to be deficient was in guerrilla fighting. With the revival of the INA in the summer of 1943 Japanese officers therefore gave the INA troops some brief guerrilla training, on the assumption that this was their major weakness as soldiers. Emphasis on guerrilla training was a universal feature of Japanese military training programs in Southeast Asia, contrasting with the training by their Western colonial predecessors. It is noteworthy in this connection that when the Burma Area Army Staff discussed the role of the INA in the Imphal campaign, Japanese officers viewed the INA chiefly as guerrilla fighters and special service units.

Bose's arrival gave impetus to the forces pushing for a campaign into India. The Imphal campaign was the only major joint campaign of Japan and the INA. Military objectives which finally dictated the launching of the offensive were the protection of north Burma from Allied counter-attack and continuing disruption of supply routes from India and Burma to Chungking. The ideà of invading India had been rejected in Tokyo in 1942. Even in late 1943 when "Operation U" was in the planning stages, there was little enthusiasm in Tokyo or in Burma. What finally convinced Tōjō were two non-military considerations. First, it was felt that one victory was essential somewhere, somehow, to improve poor morale at home in Japan. The second political consideration was Bose. Bose insisted that a joint Japanese-INA offensive be launched into India to ignite the Indian revolution. Bose's presence also allayed the fears of some staff officers that the advance of Japanese troops onto Indian soil would immediately antagonize Indian opinion. Several Japanese staff officers were impressed with Bose's undisputed leadership of the independence movement outside India. It was this personal leadership that accounted for what bargaining power Bose had *vis á vis* the Japanese. Japanese commanding generals in Burma, Kawabe of the Burma Area Army and Mutaguchi of the 15th Army, proposed that the INA be used primarily for guerrilla fighting and for intelligence and other special services, attached to Japanese units. Kawabe felt the INA would be unable to resist the temptation to desert to the British in India. On 24 January 1944, Colonel Katakura, staff officer of the Burma Area Army, met with Bose to outline the strategy of the offensive and the role of the INA in it. Bose, however, insisted that the INA fight in units of regimental size under Indian commanders, and that the INA spearhead the thrust across the border. Bose's argument was that the INA crossing the border would set India afire with revolution. A compromise was effected whereby the INA remained ultimately under Japanese command but in units directly under Indian officers.[10] The BAA also felt the best use of the INA politically would be in a thrust through Akyab into Bengal, where Bose was likely to meet an enthusiastic reception. But for strategic reasons it would be safer to use the INA in north Burma in the Chin Hills region.[11]

[10] Gaimushō, Ajiya kyoku, *Subasu Chandora Bosu to Nihon* (Subhas Chandra Bose and Japan), p.176.

[11] Interview with General Katakura 13 July 1966, Tokyo.

Liaison problems during the offensive accentuated acute tactical weaknesses and reflected the basic disjunction in goals. For Bose, there was the single goal of liberating India, while for Japan, Imphal was a limited holding operation subordinate to high-priority campaigns in the Pacific where supplies were inadequate. The two positions were basically irreconcilable.

The Imphal campaign was a fiasco from the time of the protracted decision to launch it to the equally painful decision to halt what was one of the worst disasters of the war for Japan. The drama began with the attempt to reach a decision to launch "Operation U". The plan was postponed twice in the planning stages. It was a badly timed and ill conceived campaign, as some in Tokyo feared from the start. The apprehensions within IGHQ were borne out in the course of events. The campaign was late in starting and ran into torrential monsoon rains. Orders of 15th Army Commander Mutaguchi to his divisional commanders could not be carried out. There were no supplies of any kind. There was no air cover for ground action or to counter Allied air power. Because of the magnitude and compounding of error, five generals were dismissed or transferred during the Imphal campaign. The results, which were to have raised morale at home with one glorious victory in the midst of defeats everywhere, had to be concealed from the populace at home.[12]

While the Japanese Army was mired in the mud and confusion of Imphal, Bose increased his demands for greater material support and supplies. He asked Kawabe to agree to enlarge the INA and to make greater efforts for victory. But Japan was already exhausted economically, and the supply problem was only one of many miscalculations at Imphal. The Japanese had no air cover at all and no anti-tank weapons. The campaign dragged out, with the command psychologically stalemated in reaching the decision to call a retreat. At a critical meeting between Generals Kawabe and Mutaguchi, each waited hopelessly for the other to intone the inevitable. Neither man spoke. While it became apparent within three weeks that the offensive would fail, it was five

[12] See account in Lebra, *Jungle Alliance,* chs. 9 and 10.

months before the order to retreat came from Tokyo.

Even after the belated order to retreat came from Tokyo on 8 July, Bose told Kawabe the INA would never retreat or stop fighting: "Even if the whole army becomes only spirit we will not stop advancing towards our homeland," Bose proclaimed. Bose's revolutionary faith once again impressed Kawabe.[13]

By the time the Imphal campaign became a fiasco, the opportunity for further effective action by the INA (and also the Japanese Army) had evaporated. The logic of the Japanese and Indian goals in Southeast Asia dictated some form of cooperation, but the limitations of the alliance were painfully revealed during the Imphal campaign. This phase of the INA-Japanese cooperation was far more complex than the almost idyllic early stage when Fujiwara and Mohan Singh had ushered the INA into being.

Was the INA a puppet or a genuine revolutionary army? The question is partly subjective and has several dimensions. Was the INA an independent army in Japanese intent, in international law and in INA aspiration?

First, the problem of Japanese intent is itself complex. There was no monolithic Japanese view of either India or the INA. Policy was formulated and implemented at several different levels, and at each command level it was colored and transformed by the biases, experiences, personalities and political predilections of the men in charge. Japanese policy did not develop as an ideal model on the desk of a staff officer in Tokyo. There were many agencies and men who, in implementing policy, in turn recreated and transformed it. The *F Kikan* is a case in point. Assigned originally to a small-scale intelligence mission in Bangkok, Fujiwara became midwife of the INA. His proposals regarding Japan's India policy got a hearing with Tōjō and Sugiyama and forced Tokyo to look west of Burma on the map.

Japan's India policy also evolved chronologically throughout the war through the pressure of factors external to the INA. Japanese attitudes at any given moment were affected by the dictates of military necessity. The *Iwakuro Kikan*

[13] Unpublished diary of General Kawabe, 22 June 1944, p.108; 12 July, p.118.

differed in character from the *F Kikan,* and the *Hikari Kikan* in turn differed from the Iwakuro organization. It was not only the men on both sides that spelled the difference. Fujiwara in 1944 would have been forced to play his role somewhat differently from the way he played it in late 1941–early 1942, regardless of his romantic idealism and genuine sympathy for Indian independence.

Second, were the Free India Provisional Government and INA independent from the standpoint of international law? Here too the answer is mixed. This question was a focal point in the court martial of INA officers on charges of treason in Delhi's Red Fort at the end of the war. If the Free India Provisional Government and its army were not independent but subordinate to Japan and the Japanese Army, then the Indians who led and participated in the FIPG and INA were legally traitors to the British Indian Government. If, on the other hand, the FIPG and INA were legally independent of the Japanese, then the officers could not be convicted as traitors, because they were leaders of an independent government in exile and revolutionary army. These were the arguments of the prosecution and defense at the post-war trials in the Red Fort.

Japan's intent and Indian aspirations are relevant here. Three separate actions toward the FIPG throw some light on Japan's wartime posture regarding the independence of the FIPG. Two days after the announcement of the formulation of the FIPG on 21 October 1943, the Japanese Government proclaimed its recognition of the nascent Indian government. But this was recognition of a provisional government, and in the view of highly placed generals did not constitute full recognition.[14]

A second action immediately followed the first. It was the announcement by Tōjō on 6 November 1943 of the transfer of the Andaman and Nicobar Islands to the FIPG. The announcement was timed to coincide with the Greater East Asia Conference in Tokyo. The FIPG now had a recognized government and territory, at least nominally. The Andaman Islands were of symbolic significance, too, as a place of political exile for Indian prisoners under the British. Though

[14] Gaimushō, *Subasu Chandora Bosu to Nihon,* p.124.

Above : Subhas Chandra Bose

Below : P.K. Sahgal, Shah Nawaz Khan and G.S. Dhillon after the Red Fort Trial

Major Fujiwara Iwaichi

Fujiwara and Mohan Singh Shaking hands

the FIPG was represented in the Islands by a commissioner, in actuality civil and military control of the Islands remained under the Japanese Navy. Bose's impatience with this situation had little effect on reality.

The third step was the appointment of a Japanese diplomatic envoy to the FIPG, a step much sought by Bose in 1944. Sending an envoy without official accreditation was another attempt to placate Bose. With these three actions the Japanese Government attempted to meet Bose's demands by granting them symbolically but not in substance.

At the INA court martial in Delhi after the war several Japanese witnesses for the defense testified. Contrary to the above wartime indications of Japanese intent, the witnesses unanimously attested that the INA was the independent military arm of an independent government in exile. The Japanese stance in 1946 was really a separate phenomenon from Japanese aims during the war. In 1946 Japanese witnesses had no wish to see leaders of the independence movement convicted by British colonial power.

Was the INA then a genuine revolutionary army? This question hinges partly on the subjective emotions of the officers and men of the INA. No one can dispute the character of Bose as a revolutionary in every sense of the word. From early school days he harbored a hatred of British rule which was accentuated rather than softened during his years in England. His refusal to assume a post in the ICS which he won through examination was a significant step in the metamorphosis of Bose the revolutionary. For Bose there could be no cooperation with the imperialist power. His conviction that the only way to rid India of British rule was to expel it by force was the decisive step in the formulation of Bose's revolutionary faith. It was also the point of contention between Bose and the Gandhi/Nehru axis of the mainstream of the Congress Party. But Indian revolutionary strength had to be augmented by foreign power, and Bose turned to Italy, Germany, Japan, and finally Soviet Russia in search of outside support. At the end of the war, Bose died in Taiwan in the crash of a plane bound for Manchuria in hopes of eliciting Soviet aid.

Mohan Singh, co-founder with Fujiwara of the first INA in

December 1941, was a revolutionary of a different order. Younger than Bose, Mohan Singh was a professional soldier in the British Indian Army. Until his meeting with Fujiwara in the jungles of central Malaya, Mohan Singh had rarely had a political thought. Fujiwara was the effective catalyst through which Mohan Singh came to articulate his accumulated unconscious hostilities toward the British. Mohan Singh became a revolutionary as Fujiwara watched. Mohan Singh's unwillingness to compromise with the Japanese when other Indians advised caution and moderation resulted in his arrest and confinement by the Japanese.

What of the other INA officers? Most of them, including Mohan Singh, felt a conflict of loyalty when first confronted with the prospect of fighting the British Army, in co-operation with the Japanese. They were all professionals, most of them from families with traditions of long and loyal service to the British Indian Army, conditioning on which they could not turn their backs overnight. For some of these men it was several months before they could resolve these personal conflicts and volunteer for the INA. Some were convinced by the arrival of Bose in Asia. Others felt volunteering for the INA was the only way to protect Indian lives and property. There was also some careerism and even opportunism in the INA, for as volunteers they received better treatment than as POWs. One critic of INA motives asks, "They never fought the British in India. Why consider them great patriots just because they joined the Japanese in Southeast Asia?"[15] On the other hand, a Japanese wartime correspondent defends their motives, saying, "There was some professionalism, yes, but once they were near the border of India everyone in the INA was fighting for Indian freedom."[16] And Gandhi and Nehru, though they broke with Bose in 1938, conceded publicly in 1947 that he and the officers on trial were true patriots. The allegiance of Gandhi and Nehru in 1947 was to Indian national independence, not to the British Raj, and the need for national unity overcame their earlier antipathy to Bose's tactics. The answer to the original question is thus mixed.

[15] Interview with Kusum Nair, 25 January 1966, New Delhi.
[16] Interview with Maruyama Shizuo, 28 July 1967, Tokyo.

On the Japanese side, for many staff officers in IGHQ, particularly in the Operations Bureau, and for some staff officers in the field, the INA was a puppet army to be used for propaganda purposes according to Japanese requirements. For others, like Sugiyama and Arisue (Second Bureau chief), the INA might be a revolutionary army so far as the Indians were concerned, but it had to be subordinated to Japanese military and political requirements. For still others, mostly romantic idealists in the field led by Fujiwara, the INA was a genuine revolutionary army which should receive real and sympathetic support from Japan in its fight for independence form colonial rule.

Finally, the logic of geography and the common enemy made some form of cooperation between Japan and the Indian independence movement natural. Although Japan's wartime policy toward India and the INA was a peripheral concern, it nevertheless drew Japan into ever-increasing involvement. Japan's interest in the Indian independence struggle, beginning as a small-scale intelligence mission in Thailand and Malaya, developed into a complex propaganda and espionage network designed to foster anti-British sentiment, and finally burgeoned into limited support of a government in exile and revolutionary army. Despite the military defeat of Japan and with it the INA, popular support for the INA finally helped precipitate British withdrawal from India.

Compared to other Japanese-trained or fostered independence armies in Southeast Asia the Japanese military obviously held the INA in higher esteem. The proof was that the INA was the only such army with which Japanese staff officers agreed to fight in a joint campaign, albeit reluctantly. There were several reasons the Japanese finally agreed that the INA but not the BIA should fight. The INA were already experienced, disciplined fighters. Respect for British military discipline had a long tradition in the Japanese Army and particularly Navy. Another factor was the leadership of Bose. Even though he had not gone through the military training of the British Indian Army and held no formal rank he was nominal commander and undisputed leader of the INA. He had a combined political-military impact on Japanese officers.

The anomalous fact is that, though the INA was, for Japan, basically a political liaison problem, the INA was the only Japanese-fostered army in Southeast Asia to fight jointly with Japanese troops in a major campaign.

The history of Japanese cooperation with the INA differs significantly from the story of Japanese training of independence and volunteer armies elsewhere in Southeast Asia. The difference stems both from the nature of the INA and the character of Japanese goals in cooperating with the INA. The INA was formed of experienced soldiers, POWs with a long military tradition behind them. It was not a question of creating and training an army from scratch, as with the BIA, *Peta,* and the *Giyūgun.* From the standpoint of Japanese aims too, there was an important difference, though partly one of degree. Since Japanese goals in fostering the INA were part of propaganda warfare and intelligence, Japanese cooperation with the INA assumed the character of military liaison and political and diplomatic negotiations. This was true to a lesser degree of Japanese relations with the BIA, with *Peta,* and with Burmese and Indonesian leadership. But the difference in the character of the INA and Japanese goals meant that the story of the INA was a different chapter from the BIA or *Peta.*

3

The Burma Independence Army

The nucleus of the Burma Independence Army was formed even earlier than the creation of the Indian National Army. The organization of the BIA, like the INA, came about partly as a result of policy in Tokyo, partly through the initiative of the head of an intelligence agency assigned to Burma, and partly through the cooperation of young Burman nationalists.

In the case of the BIA, unlike the other Southeast Asian independence and volunteer armies, the Japanese Navy also played a role in the initial stages. The motive force generating the BIA was the outcome of both Army/Navy cooperation, and, paradoxically, also Army/Navy rivalry. The Navy was without a sphere of influence comparable to Manchukuo or China and felt nearly eclipsed by the power of the Army in the 1930's. On 7 August 1936, the Navy therefore insisted on incorporating the "Southward Advance Doctrine" (*Nanshinron*) into the "Fundamental Principles of National Policy" decided at that Conference session.[1] High-level policy statements by the Navy in 1936 regarding Southeast Asia therefore reflected not only a fundamental concern with natural resources of that region but also the Navy's search for a sphere in its rivalry with the Army.

Several individuals acted as intermediaries between the Navy and Navy-funded or encouraged research groups. Notable among them was one Kokubu Shōzō, an ex-Navy man who had spent some nineteen years living in Burma and was especially conversant with Burmese nationalism. Kokubu had been discharged from the Navy as a captain, but he still maintained contacts with influential former classmates. He

[1] Gaimushō, *Nihon gaikō nempyō narabi ni shūyo monjo* (Chronology of Japanese diplomacy together with important documents)v.II, p.344.

plied his friends in the Navy with reports and articles and published a two-volume study on Burma during the war.[2]

Among the numerous private and semi-official research institutes in Japan concerned with Southeast Asia was the *Nan'yō Kyōkai* (South Seas Association), whose leading spirit was another retired Navy man, Konishi Takehiko, also a former classmate of Kokubu. An influential Deputy Minister of Foreign Affairs, Ōhashi Chūichi, was alleged to have helped the group, perhaps even financially.[3]

When Kokubu returned from Burma to Tokyo in 1940 a second association was formed relating specifically to Burma, the *Biruma Kenkyūkai,* and many members of the *Nan'yo Kyōkai* also joined the new group. Kokubu elicited help from his powerful former classmates, now highly placed in the Navy, including Admirals Hoshina Zenshirō and Kusaka Ryūnosuke, and Vice-Admiral Maeda Minoru. There are indications that the *Biruma Kenkyūkai* had also received funds secretly through the Navy.[4] Kokubu and Konishi cooperated in this association, whose main purpose was to foster closer relations between Japan and Burma, and more specifically, to aid the Burmese independence movement. There was thus a kind of Burma lobby operating in Tokyo to influence Japanese policy and interests in Burma by 1940, which was not true for India or Indonesia.

Among the officially sanctioned research organizations conducting studies and issuing position papers in 1940 and 1941 were the National Policy Research Institute and the Total War Research Institute. To what degree their proposals influenced government policy is a moot question. As early as 20 September 1940, however, the Nationality Question Committee of the National Policy Research Institute issued a secret report titled "Measures to be taken toward the Peoples of East Asia – Measures for Burma."[5] According to this report "The purpose is to free Burma, as part of the Greater East Asia Co-Prosperity Sphere, from the fetters of British

[2] Kokubu Shōzō, *Dai Biruma shi* (History of Greater Burma), 2 vols., Tokyo, 1944.
[3] Private interviews, Tokyo and Kagoshima, 1970.
[4] *Ibid.*
[5] International Military Tribunal for the Far East, Exhibit 1029.

imperialism as soon as possible." The National Policy Research Institute thus posited two major themes of Japan's Burma policy: 1) that Burma should be included in the Greater East Asia Co-Prosperity Sphere, and 2) that Burma should be liberated. Neither of these two tenets was fundamental to Army strategic planning when in fact the Army prepared to invade Burma at the beginning of 1942. But a Foreign Ministry "Outline of Japanese Foreign Policy" dated 28 September 1940 also assumed that Burma would be incorporated into the Greater East Asia Co-Prosperity Sphere.[6] This assumption was re-enforced by a joint Army-Navy proposal forwarded to IGHQ on 26 July 1940, titled "Outline of the Policy to cope with the World Situation" and adopted as a Liaison Conference decision. It envisioned a self-sufficient economic structure based on a nucleus composed of Japan, Manchukuo and China, with the incorporation of the Southern Area east of India, and north of Australia and New Zealand.[7]

A War Ministry policy review in 1941 saw it as necessary for Japan to occupy Burma for two reasons: to establish right-flank key positions for the Japanese defense line against the enemy in the Indian Ocean front, and to effect the capitulation of the Chungking regime by disrupting the Burma Road to Yunnan. It was further judged that failure to capture Burma would result in the loss of the defense line and unrest in Thailand. But this position paper conceived of only limited occupation of part of southern Burma initially, and later capture of strategic positions as the war situation required.[8]

[6] International Military Tribunal for the Far East, Exhibit 628, Doct. 327-A. Professor Won Z. Yoon also emphasizes Japan's goal of liberating Burma in *Japan's Scheme for the Liberation of Burma: The Role of the Minami Kikan and the Thirty Comrades.* Japanese policy toward Burmese independence was actually more ambivalent than Professor Yoon would have us believe.

[7] Hattori Takushiro, *Complete history of the Greater East Asia War,* translation, v.I, p. 35.

[8] Hattori Takushirō, *Dai Tōa sensō zenshi* (Complete History of the Greater East Asia War), p.266. Ba Maw reports being approached by the Japanese when he was premier in 1938 with an offer of a substantial sum of money if the Burma Road could somehow be immobilized: *Breakthrough in Burma,* p.106.

When strategy was later drafted for launching the invasion of Burma, another imperative was added. Politically the occupation of Burma would accelerate the alienation of Burma from Great Britain. This had in fact led IGHQ initially to consider operations against all of Burma. It was only in face of Army reassessment of priorities just prior to the outbreak of war that Burma objectives were curtailed. The Southern Army was therefore ordered to capture the air bases in southern Burma when the opportunity arose and, if the situation permitted during a pause in the Southern Invasion operation, to conduct an offensive for all of Burma. But even as late as the outbreak of hostilities in the Pacific there was no consensus in Tokyo on the best way to defend Burma.[9]

The operational plan transmitted by IGHQ to General Iida Shōjirō, Commander of the 15th Army, envisioned a three-stage operation against Burma: an initial stage in which enemy air bases in the south would be destroyed; a second stage when the Rangoon sector would be attacked along with key bases of Chinese-British operations, and a final stage when the 15th Army, with reinforcements, would finally annihilate British and Chinese forces in Burma.[10]

When the 15th Army accordingly crossed the Thai-Burma border on 22 January 1942 IGHQ revised its estimate in favor of an immediate, full-scale offensive in Burma. A new Burma Operation order declared: "The objective of the Burma Offensive is to occupy and secure the strategic areas in Burma by destroying the British Army in Burma and, at the same time, intensify the blockade against China."[11] The Southern Army Headquarters accordingly transmitted to the 15th Army an order on 9 February to follow through the thrust to Rangoon and gain a foothold as far north as possible, preparing for operations against Mandalay and Yenangyaung.[12] For a while the Southern Army had opposed the Tokyo plan to carry out a full-scale campaign against Burma, preferring to concentrate on the Burma Road and

9 Hattori, *Dai Tōa sensō zenshi,* p. 266.
10 *Ibid.*
11 *Ibid.*, p.267.
12 *Ibid.*, pp.267-8.

strategic points near Rangoon. Tokyo overrode the Southern Army's opposition, and it was forced to transmit the orders to General Iida.[13]

When Burma entered the arena of strategic military planning the earlier policy proposals on Burma were modified and goals assumed more modest proportions. The Total War Research Institute, in its 18 February 1942 report titled "Establishment of East Asia; Maneuvers for the First Period of Total War", stated that "Strict military administration will be established in Burma as it is expected to be adjacent to the front for quite a long period. However, the existence of the Burmese own administrative organ will be recognized and this under our guidance will become the nucleus of an independent government in the future." This proclamation in fact was closer to the policy pursued by Japan in Burma than earlier statements and proposals regarding independence for Burma.[14]

There were many disjunctions and ambiguities in Japan's policy towards Burma and Burmese independence between 1940 and 1942. Their prevalence in fact leads to the conclusion that Japan had no coherent Burma policy. In Tokyo, there were many pronouncements that Japan's goal in Burma was to encourage independence, but these were at times virtually cancelled by contrary statements, often nearly simultaneously. There were similar lateral disagreements with Southern Army Headquarters and within 15th Army Staff Headquarters on dealing with the Burmese independence movement. Further disjunctions appeared vertically between levels in the command structure. Between IGHQ and the Southern Army, between the Southern Army and the 15th Army, between the 15th Army and the *Minami Kikan*, Burmese independence was a thorny issue.[15]

The Japanese Army was on the whole more enthusiastic

[13] Imai Takeo, "Ajiya dokuritsu ni hatashita meishu Nihongun no kozai", (The merits and demerits of Japanese Army leadership in achieving Asian independence), *Maru,* v.XX, no.9, Sept. 1967, pp.212-9.

[14] International Military Tribunal for the Far East, Exhibit 1335.

[15] See discussion on dualism in Japan's Burma policy in Ohta Tsunezo, "Japanese Military Occupation of Burma – the Dichotomy", *Intisari,* II, 3, pp.25-43.

regarding the Indian independence movement than for aiding Burmese independence. This difference derived in part from the fundamental fact that the Japanese Army occupied Burma but not India. General Iida Shōjirō of the 15th Army, frustrated at not being able to affect decisions made by IGHQ and the Southern Army, regretted later that "Burmese independence was only used as a means for carrying out Japan's war."[16] General Iida professed sympathy for Burmese aspirations and noted, "We were not allowed to utter even the first letter 'I' of the word 'independence'." Iida felt there was a danger, if Japan avoided the word "independence", of conveying the impression that Japan had simply replaced England as the colonial power in Burma.[17]

India was the stronghold of British power in Asia, but Japan was unable to adopt a military solution to the problem of British power in India. In the case of Burma, however, Japan did seek a military solution. Burma was a battlefield while India was not, or became so only briefly and abortively. The goal of promoting Burmese independence, insofar as it was articulated, was often seen as ancillary to the higher priority goal of stimulating the Indian independence movement.

Confusion in Tokyo regarding policy towards Burma and the Burmese independence movement is reflected in two nearly simultaneous pronouncements by the same policy-making body: the Imperial General Headquarters-Government Liaison Conference (the service ministers, chiefs of staff, Foreign and Prime Ministers). In a decision of the 69th Liaison Conference on 15 November 1941, it was stated that "the independence of Burma will be promoted, and this will be used to stimulate the independence of India". Five days later, however, on 20 November, it was decided "to lead and encourage the native peoples to have deep appreciation and trust for the Imperial Army, and to avoid any action that may stimulate unduly or induce an early independence movement".[18]

16 Iida Shōjirō, *Senjin yūwa* (Twilight battlefield tales), p.74.
17 Imai Takeo, "Ajiya dokuritsu", p.217; Bōeichō kenshūjo senshishitsu, *Biruma kōryaku sakusen* (Burma offensive operation), p.519.
18 Ike, Nobutaka, *Japan's Decision for War, Records of the 1941 Policy Conferences,* pp.248, 252. See also Professor Ike's discussion on the origins and functions of the Liaison Conference.

Similar confusion and contradiction existed within Southern Army Headquarters, especially after the arrival of General Kuroda as Chief of Staff in Singapore, in July 1942. Kuroda, perhaps influenced by sympathy in Tokyo for Burmese independence at an early date, insisted on recognizing independence. He was vehemently opposed within Southern Army Headquarters by Colonels Ishii Akiho, Chief of the 3rd Section, and Ishii Masami, Chief of the 1st Section of Staff Headquarters, who felt it was imperative to give priority to operations. They felt to give Burma independence in the midst of war would be disruptive of operations and might further lead to the same mistakes Japan had made with Wang Ch'ing-wei in Nanking. There was also the problem of choosing reliable leadership among the competing political factions in Burma.[19] This conflict was ultimately resolved not within Southern Army Headquarters but by decisions in Tokyo on the timing of Burmese independence in 1943.

Confusion also reigned within 15th Army Headquarters in Burma once the Army entered Burma regarding how to deal with the independence movement. General Iida's aspirations for Burma were reflected in his 22 January appeal to the Burmese people for cooperation. General Iida announced: "The aim of the Burmese advance of the Japanese Army is to sweep away British power which has been exploiting and oppressing you for a hundred years, and to liberate all Burmese people and support your aspirations for independence."[20] Iida invoked Tōjō's statement on 21 January that Japan would grant independence if the people would cooperate with Japan. By 15 March 1942, however, the 15th Army Staff, subject to the imprecations of the Southern Army and Tokyo, and because of disagreements within the 15th Army command structure regarding independence, had become much more cautious. According to Article 2 of the

[19] Colonel Ishii Akiho, *Gunsei nikki* (Diary of military government), quoted in *Biruma kōryaku sakusen*, pp.518-19; Sawamoto Rikichirō, *Nihon de mita Birumagun no seiritsu* (The establishment of the Burmese Army viewed from Japan), v.II, pp.238-9, unpublished.

[20] Ōta Tsunezō, *Biruma ni okeru Nihon gunseishi no kenkyū* (Studies of Japanese military administration in Burma), pp.45-6; Imai, "Ajiya dokuritsu", p.218.

Hayashi Shūdan Senryōchi Tōji Yōkō (Outline of the Hayashi Group Occupational Administration), "Establishment of an independent Burmese government may be considered only after the conclusion of the Pacific War. Therefore for the time being decisions on this issue are to be suspended. However, careful guidance and consideration must be given to the people so that they may not lose hope for the future."[21] The basic decision to establish military administration in Burma was made in Tokyo by December 1941, on the outbreak of hostilities.[22]

The *Minami Kikan* played a critical role in the creation of the Burma Independence Army. Initiative in the formation of the *Minami Kikan* lay with two men, Kokubu Shōzō, and Suzuki Keiji, a Colonel in Army General Staff Headquarters. Suzuki was born to a farming family in Hamamatsu in 1897. After his graduation from the Military Academy and Staff College he became a staff officer in Headquarters. In 1932 he was sent to the Philippines as military attaché, where among other activities he charted the depth of the ocean around Manila Bay. Returning in 1935 he became instructor at the Military Academy. In 1939, he was assigned head of the Shipping Section in IGHQ. At the end of 1939, he was ordered to Djakarta to study problems of natural resources and the deployment of soldiers. But because of the Army's great concern with Chungking's source of supplies through Burma, the orders were cancelled at the last minute and Suzuki was ordered instead to Burma.

Why should the Army have given Suzuki a special assignment of this nature? He was a man of independence of thought and action. His career generally marked him as one of the élite in General Staff Headquarters. He might be ideally suited to an intelligence assignment which required him to act on his own initiative in unpredictable circumstances. The orders given Suzuki in February 1940 by the Chief of General Staff were: 1) to go to Burma to investigate the route supporting Chiang K'ai-shek with a view

21 Quoted by Ohta in "Japanese Military Occupation of Burma", *Intisari*, p.31.

22 International Military Tribunal in the Far East, Exhibit 1334, Doct. 1987-B.

to disrupting it; and 2) to study and make contact with the Burmese nationalist movement. In carrying out this dual mission he was to conceal his name and status.[23] Suzuki's activities for the next few months in China and Burma antedated the formal establishment of the *Minami Kikan* by several months.

Suzuki's swashbuckling, eccentric character had always made it difficult for him to fit himself into a bureaucratic military command structure. But his eccentricity and romanticism also were assets in an assignment in intelligence, where adaptability and ingenuity were the ingredients of success. Suzuki took the name Bo Mogyo ("Thunder-clap"), either at his own or Aung San's instance, because it identified him with a widely believed folk prophecy that lightning would one day shatter the umbrella rod (read: British rule) and save Burma. A rumour was also spread that Suzuki was the descendant of a Burmese prince who had fled from the British to Thailand. Suzuki thus helped to foster a dual legend revolving around himself which may have helped to give the activities of the *Minami Kikan* a mythical kind of legitimacy in the eyes of the Burmese people.[24] This legend was also symbolic of Suzuki's personal involvement with the cause of Burmese independence.

Suzuki's attempt to instill in his Burmese comrades a spirit of independence, even if it meant criticism of the Japanese Army, is illustrated by his often-quoted advice to U Nu: "Independence is not the kind of thing you can get by begging for it from other people. You should proclaim it yourselves. The Japanese refuse to give it? Very well then: tell them that you will cross over to some place like Twante and proclaim independence and set up your own government. What's the difficulty about that? If they start shooting, you just shoot back."[25]

Suzuki's willingness to criticize the Japanese Army was also symptomatic of the duality of his own image of his

[23] Information derived from an interview with Suzuki Keiji by Hirano Jirō, July 1962, Hamamatsu City.

[24] Ba Maw. *Breakthrough in Burma, Memoirs of a Revolution, 1939-1946*, p.139.

[25] U Nu, *Burma under the Japanese*, pp.24-5.

mission, an ambivalence which made it possible for him to countermand orders from Tokyo later. Suzuki's dedication to Burmese independence and to the creation of the Burma Independence Army were not clearly dictated by military necessity from the Japanese standpoint. These factors also contributed to the undermining of Suzuki's credibility in Tokyo, and with Southern Army and 15th Army Headquarters. And the fact that 15th Army Headquarters had not even heard of the *Minami Kikan* when in December 1941 Suzuki first contacted the 15th Army Staff made Suzuki even less inclined to pay attention to the command structure than he normally would.

In February 1940, Suzuki got his orders from IGHQ to go to Burma. He spent the next three months gathering information in Japan and China, making preparations for his assignment. It was suggested that he go in the guise of a *Dōmei* News Agency correspondent, and he accordingly requested an assignment of the chief of *Dōmei*. Suzuki was refused. A friend suggested then that he apply to *Yomiuri Shimbun*, one of Japan's largest newspapers. This time Suzuki was successful, through his *Yomiuri* friend, and he became nominally a *Yomiuri* correspondent. Another problem was a pseudonym. While paging through the Tokyo telephone directory, he came across the name Minami, "South". Since he was assigned to Southeast Asia, Minami seemed an appropriate surname; he chose Masayo as a given name, and prepared a passport and other documents under this alias. But there was little information in Army General Staff Headquarters about Burma.

Suzuki had been in Shanghai and had a valuable acquaintance with a former *tokumu kikan* man in Shanghai, one Higuchi Takeshi. Through Higuchi, Suzuki gained two co-workers who were later among the original members of the *Minami Kikan*. One of these men, Sugii Mitsuru, was a civilian graduate of Taipei Commercial University who had been for many years a businessman in Indonesia and in several parts of South America. Sugii in 1940 was serving in the *Kōain* (East Asia Development Board) in Shanghai. Higuchi approached Sugii several times before he overcame Sugii's reluctance to consider serving in Southeast Asia before

Japan had solved the problem of China. When Sugii finally met Suzuki he was amazed to find that he was someone he had known when he was at school in Taiwan.[26]

The second recruit found by Higuchi was Mizutani Inao, another civilian, then serving in the Research Bureau of the South Manchuria Railway Company. The Research Bureau was doing research not only on China but on Southeast Asia as well. Mizutani was a graduate of Tokyo University of Foreign Languages. Neither man had ever had any military training or been in Burma.

Suzuki's problem was to provide funds to attract these two men and to procure travel funds, as it was still several months before IGHQ established and funded the *Minami Kikan.* Suzuki wanted to take Sugii and Mizutani with him to Burma. Higuchi persuaded a friend, Okada Kozaburō of the Ensuiko Sugar Refinery Company in Taiwan, to provide the necessary funds.[27] Sugii and Mizutani were given cover assignments as advisers to the Ensuiko Sugar Refinery Company. Suzuki was now ready to venture to Burma as correspondent for *Yomiuri* and secretary of the Japan-Burma Society, a cultural organization to promote mutual friendship.[28]

In late May 1940, Suzuki flew to Bangkok, where he had his first encounter with Colonel Tamura Hiroshi, military attaché and head of the Shōwa Tsūsho. From there he pushed on to Rangoon, where he remained until the following January carrying out his initial mission. Sugii and Mizutani followed Suzuki to Rangoon. From the Iroha Hotel on Merchant Street in Rangoon, the three men observed the daily transhipment of arms and ammunition from ship to train for Myitkyina and from there by road to Bhamo or Mandalay and Yunnan. Suzuki and his colleagues soon learned that Burmese nationalists often attacked the trains as they moved weapons up-country. The next logical step was

26 Interview with Sugii Mitsuru, 17 November 1970, Tokyo.
27 Interview with Sugii; and Hirano interview with Suzuki, *op. cit.*
28 Sugii Mitsuru, *Minami Kikan gaishi* (Unofficial history of the *Minami Kikan),* ch.IV; Izumiya Tatsurō, *Biruma dokuritsu hishi, Sono na wa Minami Kikan* (Secret history of Burmese independence; Its name, the *Minami Kikan*), p.21.

therefore to contact Burmese nationalists and coordinate the two aspects of the mission.[29]

The problem was how to contact Burmese nationalist leaders. Suzuki made friends with several Japanese of many years' residence in Burma: Nagai, a Buddhist priest; Murakami, a Rangoon masseur; Oba, chief of the Japanese Trade Bureau; and Dr. Suzuki, an M.D. licensed in London. Besides, Col. Suzuki met his counterpart, Kokubu Shōzō, and his wife, a practising dentist. From Oba and Nagai in particular Suzuki learned something of the factions and parties in the nationalist movement. Nagai's temple and the Japan-Burma Society, Suzuki discovered, were good places to meet useful Burmans.

The Thakin Party, Suzuki learned, was the most important arm of the independence movement. It in some ways overshadowed the other two parties, the Sinyetha Party, led by Dr. Ba Maw, English-trained barrister and wartime leader, and the Myochit Party, led by U Saw, Minister of Forestry and later Premier. Suzuki learned too that Kokubu had established close connections with the Aung Than—Ba Sein faction of the Thakin Party and that Kokubu regarded Aung San and his faction as Communists unfit for Japanese support. Suzuki, however, learned that Kokubu was being watched by British authorities, and he therefore deemed it unwise to work with Kokubu's contacts.[30] There was in addition some personal antipathy between Suzuki and Kokubu, reflecting their Army and Navy affiliations respectively, which influenced Suzuki's decisions.

Suzuki was approached by U Saw offering cooperation in return for large-scale material assistance in arms and money. Suzuki told U Saw he was only a newspaper reporter and U Saw should discuss the matter at an official level, perhaps with the Japanese Consul.[31] The Consul General had also been maneuvering to make contacts with Burmese nationalists, and was a third but less potent source of initiative in such activities, apart from the Army and Navy.

29 Hirano Jirō, "A Study of the Minami Kikan", pp.40-1.
30 Interview with Takahashi Hachirō, 23 September 1970, Tokyo.
31 Suzuki Keiji, "Aung San and the Burma Independence Army", in Maung Maung, ed., *Aung San of Burma,* pp.55-6.

Above : Colonel Suzuki Keiji
Left : Lieutenant Yanagawa Munenari

Above : I. Lubis as Chief of Staff
Below : I. Lubis going on pilgrimage to Mecca

Suzuki discovered that Burmese nationalists were divided on possible sources of foreign aid, one faction looking to China and Russia, the other favouring Japan. Suzuki feared the first group was in the majority, and he wanted to shift the balance to an orientation toward Tokyo. One way to do this, he decided, and an effective way for Japan to help, was to give military training to young Burmans, as they had no military experience. Independence could not be achieved without an army. Such training obviously could not be given in British Burma, so young Burmans would have to be selected and smuggled out, first to Bangkok, then to Japan. If the Army refused to sanction the plan, Suzuki would see to it personally.[32]

In October 1939, the three major parties created an alliance, the Freedom Bloc, led by Ba Maw. By 1940, however, many nationalist leaders were either in British jails or had gone underground. Suzuki was therefore unable to meet the leaders of either faction personally at the time. During 1940, British authorities arrested members of the Thakin Party wholesale. Thakin leaders decided that foreign military assistance was essential and that the struggle must coincide with the war in Asia which was apparently imminent. Some put their hopes in China, others favored Japan, and still others sounded out Thailand.

Suzuki's arrival in Rangoon in June 1940 therefore precipitated a debate within Thakin leadership on whether to accept Japanese help. Thakins Mu, Than Tun, and Soe, who opposed cooperation with Axis Powers under any circumstances, were already interned. Those who reasoned that without foreign assistance against the British, independence would never be achieved, won out. It did not matter whether the aid came from China or Japan according to this line of argument, but it had to be armed assistance. Training with arms would make real revolution a possibility. But Japan should pledge independence, and if Japan then failed to keep her vow, Japan too would be fought until independence became a reality.[33] The rationale for armed cooperation with

32 Hirano interview with Suzuki; Suzuki, "Aung San and the BIA", p.56.

33 Ba Than, Dhammika U, *The Roots of Revolution,* pp.14-15.

Japan was thus much the same as with the leadership of the INA.

A forerunner of Japanese-Burmese friendship and pioneer of radical religious nationalism in Burma was U Ottama, a monk from Arakan who went to Japan in 1907 and again in 1912 to study Japanese Buddhism. While in Japan he taught Pali at Tokyo University and studied at Kyoto University. He was extremely impressed by Japan's modernization and the industry of the people.[34] Japan during his first visit was still exultant over the victory over Russia and was hosting nationalist students from all over Asia. Returning to Burma in 1915 from his Japanese sojourn, he organized an All-Burma Buddhist Youth Association, prototype for many later religious-political groups. Through these groups, he sought to transpose his Japanese inspiration and Buddhist precepts into the political arena. First rallying point for this religious nationalism was the ban on foreigners entering pagodas wearing shoes. U Ottama was an archetypal political monk, and for his efforts he was jailed several times for sedition. U Ottama also reflected the Asian nationalist orientation toward Japan following the Russo-Japanese War.

Most of the Thakins were former students at Rangoon University, where they were active members of the Students' Union. Aung San, one of the most radical students, was suspended following criticism of the university administration. U Nu was suspended for a similar attack. The two students then worked actively for the *Dobama Asiayone* (We Burmans Association), also known as the Thakins. The *Dobama Asiayone* was founded in 1930 by Thakins Ba Sein and Kodaw Hmaing. In 1939, a small group of Thakins led by Ba Swe and Kyaw Nyein formed a secret sub-group, the Burma Revolutionary Party.[35] Thakins Aung San and Mya also were active in this revolutionary group and discussed plans for resistance against the British. Aung San brought the plan for armed resistance to the attention of Ba Maw who realized how serious the BRP was and how useful an understanding with the Japanese might be.[36]

[34] Kokubu Shōzō, "Biruma seiji undō no tembo", (A survey of Burmese political movements), p.3.
[35] Tinker, Hugh, *Union of Burma*, pp.6-7.
[36] Ba Maw, *Breakthrough in Burma*, pp.71-4.

In 1940, Aung San attended the Ramgarh session of the Indian National Congress. Shortly thereafter he eluded a British order for his arrest and escaped to China with Hla Myaing.[37] Ba Maw recalls that it was the suggestion of Japanese vice-consul Fuki that Aung San escape to Amoy, where he could be picked up by the Japanese Army and taken to Japan.[38] There was, however, no direct connection between Fuki's suggestion and the arrangements actually made by Suzuki.

One key figure in the early stages of Japanese-Burmese liaison was Dr. Thein Maung, a colleague of Dr. Ba Maw and a man with Japanese friends in Rangoon. Thein Maung was active in the Japan-Burma Society, whose opening ceremony he attended in Tokyo in 1939. Ba Maw sent him to Tokyo at the suggestion of the Japanese Consul. In Tokyo, he spoke publicly and returned with promises of Japanese aid, according to Ba Maw.[39] Thein Maung also spent time in jail with Ba Maw. Suzuki met frequently with Thein Maung and through him contacted Thakins Mya and Kodaw Hmaing, a sixty-year-old poet respected by all Thakins.[40]

In September 1940, Thein Maung showed Suzuki pictures of Aung San and Hla Myaing, who had escaped to Amoy disguised as Chinese laborers aboard a Norwegian freighter. Thein Maung sought Suzuki's aid in rescuing the two Thakins. Suzuki took the pictures and decided this was an opportunity to get at least two Thakin leaders to Tokyo for training. On 3 October, he left Rangoon for Bangkok, where he made arrangements with Colonel Tamura for aiding the escape of young Burmans from Rangoon to Tokyo. The Japanese Embassy in Bangkok would be the rendezvous point for those escaping across the border via Chiengmai, Mesot, and Tavoy. Suzuki then flew to Taipei where he met a friend in Army Headquarters and asked him to send officers to Amoy to contact Aung San and arrange his transport to Tokyo.[41]

[37] Tinker, *Union of Burma*, pp.6-7.
[38] Ba Maw, *Breakthrough in Burma*, pp.120-2.
[39] *Ibid.*, p.63.
[40] Ba Maw, *Breakthrough in Burma*, pp.111-2.
[41] Suzuki, "Aung San and the BIA", p.56.

By January 1941 Suzuki was back at the 8th section of IGHQ in Tokyo. He received a call from the police at Hakata Airport in Kyushu who had in custody two foreigners who had flown in from Taiwan. Suzuki learned from the police description that the two were Aung San and Hla Myaing who had been brought from Amoy. Suzuki made arrangements for their accommodation in Tokyo, then took them home with him to Hamamatsu. He got to know the two men personally and gave them some preliminary training in handling swords and rifles.[42]

In mid-1940 the Navy General Staff knew more about the political and economic situation in Burma than did Army General Staff Headquarters. Kokubu Shōzō had been keeping Navy Headquarters informed on the political situation and economic resources through a series of reports based on several years' first-hand acquaintance. Army General Staff Headquarters, on the other hand, had no such source of information on Burma, but had just sent Suzuki to Burma to learn what he could.

Kokubu's main contacts in the independence movement, by contrast, were with the Aung Than and Ba Sein faction of the Thakin Party. With them Kokubu had discussed the drafting of a plan for Burma's independence. Kokubu also knew the general situation in Burma well and had taken a Foreign Ministry official, Ōhashi Chūichi, on a tour of a railway viaduct on the Burma Road. For Kokubu's pains he and Ōhashi were arrested by British police who suspected a plot to blow up the bridge. Through Kokubu's friends in both Navy Headquarters and the Foreign Ministry he was called to Naval Headquarters and asked for information.

The Navy was already sensitive to the problem of policy toward the South. Kokubu suggested a plan for a Burma policy, basing his proposals on his many conversations with Aung Than and Ba Sein and on a plan drafted by Ba Sein.[43] Kokubu's plan for Burmese independence envisaged overthrow of the pro-British U Pu cabinet (1939-40), establishing a new cabinet under U Ba Pe, who would in turn be overthrown by a non-confidence bill and replaced by Ba Maw; supplying

[42] Hirano interview with Suzuki.
[43] Ōta Tsunezō, *Biruma ni okeru Nihon gunseishi*, pp.39-40.

weapons, ammunition, communications devices; mobilizing press and priests to achieve unity of these projects to be supported by funds from Japan; and minting of Burmese currency notes in Japan. Kokubu conceived a plan for Burmese units to be supplied with arms, though he did not detail a program for military training. U Ba Pe cabinet members, Ba Maw's party and part of the Thakin Party, the General Council of Buddhist Associations, and groups of young priests would all be mobilized for political and military action. A Japanese headquarters would be established in Rangoon, with branches in Moulmein, Bassein and Mandalay.[44] To create forces capable of military action, Kokubu suggested mobilizing farmers near Mandalay, Moulmein, Bassein and the outskirts of Rangoon. He spoke of a revolutionary army but not of the details of its recruitment or training. He also suggested a "revolutionary provisional government", dissolution of parliament when military occupation was completed, and establishing a semi-military government under martial law. Kokubu believed the Burmese people were so nationalistic and pro-Japanese that he dismissed the need for propaganda groups in Burma.[45]

Kokubu tried to impress on Navy General Headquarters the advisability of supporting Aung Than and Ba Sein rather than Aung San, who he was convinced was a Communist. This led to friction between Kokubu and Suzuki as champions of rival factions of the Thakin Party. Despite Kokubu's much longer first-hand acquaintance with the Burmese nationalists, he did not prevail over Suzuki and the more powerful Army in this initial difference of opinion.

With Kokubu and Suzuki feeding information to Navy and Army Headquarters respectively, and both in Tokyo in January 1941, it seemed to both the Army and Navy that the time had come for concrete steps. Aung San's presence in Japan may have been an added incentive. The Navy contacted the Army, and this precipitated joint conversations leading to the formation of the *Minami Kikan.* On 16

[44] Kokubu Shōzō, "Independence Draft Plan", in Gaimushō, *Dai Toā sensō kankei ikken, Biruma mondai* (Matters concerning the Greater East Asia War; the Burma problem), n.d., p.370.
[45] *Ibid.*

January, the first of several meetings was held at the *Suikosha* (Navy Club). Suzuki and one of the leading Army members, Lt. Kawashima Takenobu, prepared a draft plan on Burmese independence, apart from the Kokubu draft. Sugii and Mizutani were involved in the preparations as civilians from the Army side. It was decided to give the *Kikan* the cover name of *Nampō Kigyō Chōsakai* (Southern Enterprises Research Association).[46]

Minami Kikan members included ten officers who were graduates of the *Nakano Gakkō* and over fifteen NCOs who were also graduates. Though the NCOs were not trained in intelligence, they had studied the Burmese language as part of their preparation.[47] Most of the *Nakano Gakkō* alumni were graduates of the second class and also graduates of the 46th through 49th classes of the Military Academy.[48]

On 1 February 1941, the *Minami Kikan* was brought to life as a joint Army-Navy venture, under Colonel Suzuki Keiji. Its chief aim was to cut the Burma route supplying Chungking; an ancillary assignment was to aid Burmese independence. Army and Navy members met several times during February to discuss plans and tactics. The basic plan was to bring thirty Thakin Party members out of Burma, train them, supply them with arms and ammunition, and smuggle them back into Burma, where they would instigate anti-British military action. Southern Burma would be occupied by guerrilla forces and an independent government would be created.[49] There is some question as to whether IGHQ approved the last point.

Passports were arranged under false names for Aung San and Hla Myaing, who were provided with false teeth as part of their disguise. Sugii was instructed to arrange with the Daido Steamship Company for smuggling Aung San into Burma by ship and was to accompany him. Sugii was also to take charge of smuggling other members out of Burma and

[46] Sugii, *Minami Kikan gaishi,* ch.VIII.
[47] Interview with Kawashima Takenobu, 3 July 1972, Tokyo.
[48] Interview with Izumiya Tatsurō, 21 October 1970, Tokyo; Maruyama Shizuo, *Nakano Gakkō*, p.3.
[49] Ōta, *Biruma ni okeru Nihon gunseishi,* pp.39-40; Sugii, *Minami Kikan gaishi,* ch.VIII; Izumiya, *Minami Kikan,* p.36.

transporting them to their training camp on Hainan Island. Communications with IGHQ were to be through Capt. Ozeki of Army General Staff Headquarters and Commander Hidaka of Naval Headquarters.[50] Navy Captain Ono in the Consulate General in Rangoon was to be in charge of liaison with Thakin Party members and to assist in their exodus from Burma. Reports were to be filed with IGHQ or with Col. Tamura in Bangkok.[51] Aung San with Sugii and Hla Myaing with Mizutani were then smuggled back to Burma where they collected colleagues for the return voyage to Japan. In this way, twenty-nine Thakin Party members reached Tokyo, where a Burmese student joined them to bring the number to thirty, the original "thirty comrades". Once the thirty had assembled in Japan, they were given a little preliminary training and briefing and visited the major national shrines.

The "thirty comrades" were sent from Tokyo to a special training camp near the Naval Training Center at Samah, Hainan Island. The Navy provided the center, the supplies, and arms and ammunition. The Army sent instructors to carry out the training. Training was under the direction of Captain Kawashima Takenobu, graduate of the Military Academy and the *Nakano Gakkō*. The instructors wore Navy uniforms as a disguise, but regular Navy officers and local farmers were forbidden to enter the training camp. The basic training, which began in May with eighteen Burmans, was originally scheduled as a one-month crash training program. But the training was prolonged, and more Burmans were brought from Tokyo in June. The second group was given accelerated training, including night drill, in order to catch up with the first group.[52] The language of communication between Japanese officers and Burman trainees was broken English. This was the first exposure to discipline of any kind for the thirty, and they were exhausted and demoralized at first by the rigors of Japanese-style military discipline. Still, the strong political motivation of the Thakins led them to accept the strictures of the training as a discipline to equip

50 Sugii, *Minami Kikan gaishi,* chs.IX, X.
51 Sugii, *Minami Kikan gaishi,* ch.X; Izumiya, *Minami Kikan,* pp.38-9.
52 Izumiya, *Minami Kikan,* pp.59-66; Sugii, *Minami Kikan gaishi,* ch.XIV.

them for independence. In addition to military skills and tactics, they carried out war games and were given lectures to strengthen their spirit of self-sacrifice. All training was done with loaded weapons, and special maneuvers were used to develop personal courage. Training included actual combat techniques with captured weapons, including rifles, machine guns, hand grenades and howitzers.

The Burmans were divided into three groups, with a different training emphasis in each. The first group was given regular training in command of soldiers. The second group was given special training in guerrilla tactics and espionage, while the third group was specially trained as leaders of the independence movement by Captain Kawashima personally. This latter group included Aung San and Ne Win.[53]

Kawashima was involved in all stages of training of the BIA from Hainan, Taiwan, and the Mingaladon Officers School to the final training of sixty Burmans at the Military Academy in Japan. Kawashima reports no resistance against the rigors of the training regimen, since all were eager to prepare themselves for the liberation of Burma.[54]

In July, scheduled date for the end of the training program, the thirty expected to be sent back to Burma. They awaited orders. None came. The delay was caused by changes in the international situation and discussions in Tokyo about Japan's predicament. But the Burmans in their jungle training camp knew nothing of this, and they grew increasingly restive and suspicious of Japan as time passed.[55]

In July, Japan advanced into Indochina. In August, the Army and Navy reached agreement on respective spheres of influence on the continent and surrounding waters in Southeast Asia, removing Burma from the sphere of the Navy. Suzuki judged that war with Britain and the U.S. was unavoidable and that southern Burma would surely be occupied. He attempted to placate Aung San and his comrades, who felt they should brook no further delay in the independence struggle. Suzuki in September took Kawashima, Aung San and a few comrades to Tokyo for map

[53] Sawamoto, *Nihon de mita Biruma gun no seiritsu,* I, pp.57-62.
[54] Interview with Captain Kawashima, 3 July 1972, Tokyo.
[55] Izumiya, *Minami Kikan,* pp.59-70.

maneuvers. In October, the training camp was moved from Hainan to Tamazato in Taiwan, and training was continued at an old Army camp used by the Japanese Army during the Meiji period. The move was caused by the Navy's withdrawal from the *Minami Kikan* and because an intercepted communiqué of the American Military Attaché in Tokyo revealed he knew there was a special training program in progress in Hainan.[56]

In October, there were still no orders from Tokyo regarding the thirty comrades, now reduced by a death to twenty-nine. Kawashima and Suzuki decided on their own initiative to send six Burmans into Bangkok secretly and from there to Burma. IGHQ learned of the development and sent a cable to Bangkok, ordering Suzuki to prevent the Burmans from landing and to ship them back to Tokyo instead.[57] Tokyo had not yet agreed to Suzuki's timetable, though he felt there was general agreement on the plan. A cabinet change in Tokyo saw Konoye replaced as Premier by General Tōjō on 17 October. With the international situation changing daily, it did not seem to IGHQ an opportune time for the operation, and there was the further danger that Japan's war contingency plans might be revealed. But Suzuki, angry at this reversal and unwilling to disappoint the Burmans after so many months of training, sent a cable replying that the Burmans had already landed and disappeared into the jungle. There was a further exchange of telegrams between Suzuki and IGHQ, but Suzuki refused to alter his plan. Suzuki this time cabled back falsely that the four Burmans who escaped had been apprehended and imprisoned.[58]

While the thirty Burmans were being trained, Suzuki in February 1941 left Tokyo for Bangkok, center of operations for the *Nampō Kigyō Chōsakai.* Several branches were established along the Thai-Burma border by July. These branches facilitated the smuggling of men and supplies into Burma. They functioned outwardly as commercial, mining

[56] Sawamoto, *Nihon de mita Biruma gun no seiritsu,* I, pp.67-8.
[57] Sugii, *Minami Kikan gaishi,* ch.XV; Sawamoto, *Nihon de mita Biruma gun no seiritsu,* I, pp.85-90.
[58] Izumiya, *Minami Kikan,* pp.80-1.

and forestry enterprises. A branch at Chiengmai operated on the northern border, branches at Rahaeng and Mesot covered the central border, branches at Kanchanaburi focussed on the south central sector and at Ranaung on the southern border. All branches were under Bangkok headquarters and each had a staff of three or four *Kikan* members.[59]

The Chiengmai branch operated as a commercial and development enterprise and was in this way able to investigate remote routes. The Rahaeng branch was under Lt. Takahashi Hachirō, who in the guise of an agricultural and forestry engineer was able to explore roads and rivers and to ship weapons. The Kanchanaburi branch operated a general dry goods shop and used river transport. And the Ranaung branch engaged in forestry and fishing, exploring the area around Victoria Point.[60] By November all branches were actively engaged and four Burmans had successfully crossed into Burma. In Bangkok also were assembled Burmans from throughout Thailand, recruited to cross the border and join the struggle for independence. Two hundred members were recruited before the BIA even entered Burma.

Internal problems plagued the *Minami Kikan.* Almost as soon as the *Minami Kikan* was created, friction developed between Kokubu, the old Burma hand, and Suzuki. Each man regarded himself as the champion of Burmese independence and authority on Burma. Kokubu's support of Aung Than and Ba Sein as the only trustworthy leaders and Suzuki's reliance instead on Aung San and Thakin Mya were factors. There was also a personal animosity between the two men. Furthermore, the traditional Army-Navy rivalry contributed to difficulties the Army and Navy men had in working together. The Navy felt at a disadvantage in dealing with the Army, whose prestige and power had been heightened ever since the occupation of Manchuria and war in China, as well as by domestic events of the thirties. The Navy could claim no comparable role anywhere in Asia.[61]

[59] Sugii, *Minami Kikan gaishi,* ch.XII.

[60] Sawamoto, *Nihon de mita Biruma gun no seiritsu,* I, pp.80-3; Sugii, *Minami Kikan gaishi,* ch.XII.

[61] Hirano, *Minami Kikan,* pp.72-3; Takagi Sokichi, *Taiheiyō sensō to rikukaigun no kōsō* (The Pacific War and Army-Navy Rivalry) is a valuable historical survey of Army-Navy rivalry by a Navy Admiral.

Suzuki and Kokubu also had difficulties in working out Kokubu's assignments. Kokubu was after all some nineteen years out of military service and training, and it was hardly surprising that he was not a smashing success at imparting military skills to the Burman nationalists. He did not see this as his role. He, nevertheless, rather than Suzuki, had been instrumental in smuggling ten of the thirty out of Burma. Gradually most of the Navy members of the *Kikan* were either sent back to Tokyo or forcibly hospitalized by Suzuki.[62] One staff member who identified with the Navy side reports that Suzuki gave orders to have him killed, and in fact wanted all the Navy members of the *Kikan* killed.[63] On 7 August 1941 Suzuki replaced Navy Captain Kojima with Army Captain Noda as head of the Rangoon office of the *Kikan.* Finally on 11 August the Navy Minister ordered all Navy men to leave the *Minami Kikan,* and the Navy formally withdrew its participation.[64] This was the only *tokumu kikan* where the Army and Navy joined forces, and it was an uncomfortable accommodation from the start.

Suzuki's plans for the operation of his mission included training and recruitment of the Burma Independence Army, establishment of an independent Burmese political authority in the Tenasserim region before the capture of Rangoon and north Burma, capture of enemy supplies, driving British authority from Burma using the BIA, and transfer of Burmese property and enterprises to the BIA. Suzuki discussed the plan first with Colonel Nakada Seiichirō of the 15th Army Headquarters, who then passed it on to Southern Army Staff Headquarters. Colonel Ishii Akiho and others objected strongly to the idea of a BIA provisional government and to the transfer of Burmese property to the BIA.[65] Suzuki was frustrated in his initial attempt to get the sanction of Southern Army and 15th Army Headquarters to his basic plan which he felt had already been approved in Tokyo.

62 Hirano, *Minami Kikan*, p.73.
63 Interview with Kojima, 1970, Tokyo.
64 Hirano, *Minami Kikan*, p.74.
65 Bōeichō Senshishitsu, *Biruma kōryaku sakusen*, pp.118-22.

In fact, Colonel Takeda Isao of the 8th Section in Tokyo Headquarters cabled Southern Army Headquarters that they should take the lead in establishing independence at the time of the attack on Rangoon. But Southern Army Headquarters cabled back that the situation was not favorable and they could not comply. Iida also took up the issue on 21 March in a cable to Southern Army Commander General Terauchi, stating that independence should be arranged quickly. Southern Army Headquarters ignored Iida's request.[66] During part of early 1942 Southern Army Headquarters was able to disregard both Tokyo and the 15th Army.

This dispute had repercussions among the Burmans also, who began to feel they had been deceived in being promised independence when they crossed into Burma. Aung San expressed his dissatisfaction to Sugii and to Suzuki. Suzuki, attempting to placate Aung San by provoking him, said, "If I were you I would achieve independence by any means possible. Independence should not be given by others . . . But as I am Japanese I can't point a rifle at the Japanese Army. So if you think a clash between you and the Japanese Army can be avoided by promoting independence of your own free will, why don't you just stab me with this sword and become independent?" Aung San blanched, and vowed that while Suzuki was there the BIA would not rebel against the Japanese Army.[67]

The Burma Independence Army came into being formally in December 1941 with a force of over two hundred Burmese already recruited in Thailand. The BIA was commanded by Suzuki and led by the trained thirty comrades. On 31 December, Suzuki and Aung San spoke to the new BIA recruits to encourage them as they crossed over into Burma. Cars, vehicles and transportation of all descriptions, including elephants, were requisitioned for transporting troops and supplies. Clothing, food, and weapons were distributed.[68]

As war broke out, the *Minami Kikan* was transferred from command of IGHQ in Tokyo to Southern Army Headquarters. Regiments of two divisions of the 15th Army under

[66] Sawamoto, *Nihon de mita Biruma gun no seiritsu,* II, pp.237.
[67] *Ibid.*, pp.241-2.
[68] *Ibid.*, pp.121-4.

General Iida, the 33rd and 55th, were to spearhead the Burma offensive. There was no advance preparation for a Military Government Department to deal with occupied Burma. Suzuki, on the other hand, made plans to set up temporary local administrations under the BIA after occupation of Moulmein, and to establish an independent Burmese government when Rangoon was occupied.[69] Southern Army Headquarters, when it learned of Suzuki's plans, was implacably opposed.[70]

After Japan secured control of Rangoon on 7 April, Suzuki created the so-called Baho Government under Thakins Tun Oke and Ba Sein. The mission of the Baho Government was to create some semblance of order in the administrative units of the BIA scattered through Lower Burma. Actually, administrative authority of the government did not extend much beyond the bounds of the capital of Rangoon. Individual BIA government committees were more or less autonomous. When Suzuki was transferred back to Japan, the Japanese Army decided to assume tighter control, not only over the BIA but also over the Baho Government. Accordingly, on 5 May Aung San, acting on orders from General Iida, issued an order prohibiting the BIA from any future intervention in political matters, thereby abolishing the provisional BIA committees.[71]

During the BIA interregnum in Lower Burma trouble erupted between young BIA representatives and the Karens in the Delta region. Karens were regarded by the BIA as pro-British, and the BIA soldiers attempted to collect guns from them. The Karens were unwilling to accept the authority of the BIA-Baho committees. Hostilities between Karens and Burmans in the BIA resulted in communal killings in May and ended only through Japanese intervention. It was not until reorganization of the BIA as the Burma Defense Army in 1942 that some reconciliation between Karens and Burmans was effected through creation of a Karen battalion

[69] *Ibid.*, pp.130-2.
[70] *Ibid.*, p.132.
[71] Ohno Tohru, "Biruma kokugunshi", (History of the Burmese National Army), pt.2, *Tōnan Ajiya Kenkyū* (Southeast Asian Studies), v.8, no.2, 1970, pp.347-8.

in the BDA.[72] Chins and Kachins were not included in the BIA, though there were two Shans among the original thirty.

The Japanese and the British recruited from basically different ethnic groups in Burma. The colonial forces under the British had included Englishmen, Karens, Kachins, Chins, Shans and Arakanese. There were almost no Burmans recruited. The Japanese, on the other hand, avoiding those ethnic groups recruited by the British, preferred Burman recruits. In Burma as elsewhere in Japanese-trained units, a policy of avoiding segments of the population which served the colonial régimes was followed.[73]

Japanese units took Victoria Point on 14 December without any British countermove, and the offensive in Burma was gradually extended from Mesot, Kanchanaburi and Tavoy to Moulmein. On 31 January Suzuki deployed BIA units together with the 15th Army divisions on only general orders from Southern Army Headquarters to the *Minami Kikan* to participate in the Burma operation. 15th Army Headquarters for its part had no exact knowledge about the BIA and expected it to operate only to harass the enemy to the rear as part of an intelligence operation.[74]

There was thus a fundamental difference between Southern Army Headquarters and 15th Army Headquarters on the one hand and Suzuki on the other regarding the role of the BIA as the Burma campaign was launched. The BIA had no inkling of the misunderstandings between Suzuki and the Japanese Army and thought only that the invasion of Burma would be followed immediately by independence, and that Japan would recognize that independence.[75] The assumptions of the BIA that they were secure in Japanese support of Burmese independence were strengthened by a

[72] Cady, John, *A History of Modern Burma,* pp.442-3; U Ba Than, *The Roots of Revolution,* p.40. Another version of these events is that Suzuki's reprisals against Karens for the killing of a Japanese precipitated the outbreak of hostilities between Karens and Burmans.

[73] Guyot, James F., "Ethnic Segmentation in Military Organizations: Burma and Malaysia", in Kelleher, Catherine McArdle, ed., *Political-Military Systems, Comparative Perspectives* (Beverly Hills: Sage Publications, 1974), pp.27-37.

[74] Sawamoto, *Nihon de mita Biruma gun no seiritsu,* I, pp.136-8.

[75] *Ibid.,* p.149.

statement of Tōjō in the Japanese Diet on 21 January that "the aim of the Japanese advance into Burma is to liberate all Burmese people from exploitation and suppression and to support their independence".[76] Tōjō's statement was reinforced by an announcement by General Iida to the same effect, promising that Japan's aim was to drive out British power and to liberate the Burmese people, supporting their long-cherished wish for independence.[77]

There were also difficulties in coordinating the activities of the BIA with units of the Japanese Army during the campaign. Suzuki, for example, was dissatisfied with the attitude of Lt.-Gen. Takeuchi, Commander of the 55th Division, toward the BIA, and Suzuki recalled the Kitajima Unit of the BIA from the 55th Division.[78]

As the BIA pushed through Burmese villages thousands of young men rallied to the call to join the struggle for liberation. So many joined, in fact, that neither the Japanese Army nor the BIA command were able to handle the situation. The problem of the BIA was not only its unruly size – estimated as high as 200,000 within a few weeks of its arrival in Burma – but its dissatisfaction with the Japanese Military Administration's failure to grant independence. The Southern Army Staff Headquarters sent staff officer Fujiwara, former chief of the *F Kikan,* to talk to Suzuki. Suzuki complained that Japan had promised but failed to give independence, that military administration had been imposed and Ba Maw had been selected. He warned of the possibility of revolt by the 200,000-strong BIA. Fujiwara agreed it was a danger, since the BIA had been cutting Japanese electric lines, putting up posters and clashing with the *Kempeitai.* Fujiwara suggested reducing and restraining the BIA, but this only angered Suzuki.[79] The size and unruly

[76] *Ibid.*, pp.142-3.
[77] *Ibid.*, pp.151-2.
[78] *Ibid.*, p.184.
[79] The incredibly rapid growth in the size of the BIA was due to the local tradition of dacoity and of the strength of the anti-British sentiment. For an authoritative account of the nationalist foundations of the BIA see Guyot, Dorothy, *The Political Impact of the Japanese Occupation of* Burma, Yale University, Ph.D. dissertation, 1966. See also Guyot, Dorothy, "The Burma Independence Army: A Political Movement in Military Garb", in Silverstein, Josef, *Southeast Asia in World War II,* Yale University Southeast Asia Monograph Series no. 7, pp.51-7.

behavior of the BIA made it impossible for either Japanese or BIA officers to control it. Fujiwara persuaded Suzuki to accompany him to 15th Army Headquarters at Maimyo. Fujiwara was already convinced the BIA would have to be reorganized and that the *Minami Kikan* had already outlived its usefulness. At 15th Army Headquarters Chief of Staff Isayama and Deputy Chief of Staff Nasu agreed on these two points.[80]

All the accumulated animosity between Suzuki and the 15th and Southern Army Headquarters finally contributed to a 15th Army plan to dissolve the *Minami Kikan,* incorporating some of its members into the Military Administration of Burma but transferring Suzuki back to Japan. Though Iida opposed the imposition of military government in Burma initially, he apparently offered no objection to the plan to disband the *Minami Kikan* now that Burma was actually occupied and the imposition of military government was inevitable. On 10 July the *Minami Kikan* was dissolved and Suzuki was transferred, after first designating Aung San his successor as Commander of the BIA. The BIA itself was also disbanded and drastically reorganized according to a 15th Army plan in August.

Colonel Suzuki's problem was a conflict of commitment. The *Minami Kikan* as an organ of the Japanese Army was in a chain of command ascending from the 15th Army and the Southern Army to IGHQ in Tokyo. But he also felt and often expressed in action a genuine commitment to the cause of Burmese independence. This second loyalty prompted him to take actions which the Army – whether the 15th, Southern or IGHQ – could not condone. The conflict was particularly pointed when Suzuki contravened the IGHQ orders to prevent the landing of the Burmans in Bangkok and to stop all *Kikan* operations. But divergence was also apparent when, as the Japanese Army crossed the border into Burma, Suzuki insisted that Burma be proclaimed independent immediately and the administration be transferred to an independent

[80] Sawamoto, *Nihon de mita Biruma gun no seiritsu,* II, pp.293-4.

government.[81] Suzuki in this instance got some local support from 15th Army Commander Iida Shōjirō, who after the war stated he did not feel it necessary to put Burma under military administration. But Iida's protest did not move Saigon or Tokyo to alter the plan. And the 15th Army had no information about the *Minami Kikan* at the outbreak of war, but was only informed later in December.[82] Suzuki felt that Japan, or at least he, had promised the thirty that Burma would be independent once they arrived in Burma, and that the promise was binding. In Burma, however, Deputy Chief of Staff Colonel Nasu of the 15th Army ordered that the word "promise" not be used in connection with Burmese independence. The *Minami Kikan* felt this order contravened one of the purposes of the creation of the *Minami Kikan:* assisting the Burmese independence movement. Nasu was one of those who recommended the dissolution of the *Minami Kikan.*[83]

Besides Tokyo's reaction to Suzuki's personality and actions, to a large extent the *Minami Kikan* had become obsolete once the Japanese Army occupied Burma and imposed military government. Any functions of liaison which the *Minami Kikan* had performed were now assumed by the military advisory staff of the Japanese Military Administration. The case of the *F Kikan* and *Hikari Kikan* with the Indian National Army was a different matter, since Japan was not in occupation of India, and the continued usefulness of the *Hikari Kikan* as a liaison agency was demonstrable. And the conflict between the *Minami Kikan* and the 15th Army over the role to be assigned the BIA had never been resolved. Tokyo's action in abolishing the *Minami Kikan* entirely and transferring Suzuki to Hokkaido therefore also represented reimposition of central authority by IGHQ in Tokyo.

The absence of clearly delineated policies in Tokyo toward

[81] In the words of one of Suzuki's staff, "IGHQ was raped by Suzuki. He created a result that Headquarters couldn't destroy even if it destroyed the *Minami Kikan.*" Another *Minami Kikan* member used the phrase, "Suzuki cheated IGHQ": interviews in Tokyo, September and November 1970.

[82] Sugii, *Minami Kikan gaishi,* ch.XVII.

[83] *Ibid.,* ch.XXI; interview with Major-General Nasu Yoshio, 8 October 1970, Tokyo.

Southeast Asia as late as 1941 meant that the numerous research agencies in Japan and intelligence agencies on the spot were able to move in and partly fill the vacuum. Vague and conflicting goals toward Burma were typical rather than the exception. Inchoate goals toward Southeast Asia meant that the structure of the Greater East Asia Co-Prosperity Sphere was highly divergent from one country to the next. Lack of coordination between agencies and offices even within the military establishment was the rule.

Burma was a good example of ambiguity in Japanese policy in 1941. The fact that the 15th Army Commander had not even heard of the *Minami Kikan* in December when Suzuki contacted 15th Army Headquarters convinced Suzuki that he was justified in using his own initiative without much regard for army command structure. And Suzuki could point to a statement by Tōjō in the Diet to legitimize his own encouragement of Burmese independence aspirations. Southern Army Headquarters and 15th Army Headquarters were equally unhappy with Suzuki's actions and with Tokyo's propaganda pronouncements. There was a further disjunction between the intentions of Southern Army Headquarters and 15th Army Headquarters regarding the imposition of military government in Burma. Southern Army Headquarters had least enthusiasm within the command structure for Burmese independence and the BIA.

It was partly because of this lack of a coherent Burma policy in Tokyo that Suzuki and the *Minami Kikan* were able to create the BIA, which developed a viability of its own. Meanwhile, all the cumulative problems of the BIA led to its dissolution in early 1942 and recreation as the Burma Defense Army. In 15th Army Headquarters at Maimyo Fujiwara met with Isayama, Nasu, and Takeshita, all staff officers. Takeshita devised a two-pronged plan for reorganization of the BIA: selection of a smaller, disciplined cadre of three to five thousand men as the nucleus for the reorganized force, and establishment of a staff training school at Mingaladon. Suzuki's transfer to Japan was implicit in the plan. The BDA was to be transferred from command of the Southern Army Headquarters to 15th Army Headquarters. The *Minami Kikan* was also dissolved as a result of these planning sessions.

Accordingly, by an order of Aung San dissolving the BIA on 8 July BIA members were given the option of retiring with a small pension or re-applying for enlistment in the BDA. Applicants were ordered to assemble at Mandalay, Pegu, or Mingaladon, where some 2,800 from seven or eight thousand applicants were selected by examination. Those who were admitted were between nineteen and twenty-three with at least an elementary school education. The new BDA came into being on 27 July 1942 under the command of Aung San. Three batallions were formed under Ne Win, Yan Naing, and Ze Ya, all from the original thirty. All thirty became officers of the reformed BDA. Three additional battalions were subsequently organized and stationed in Pyinmana, making a total of six infantry batallions. Finally, a seventh batallion was added. For the battalions recruited after the initial three, there was no educational qualification, and the age limits were extended from eighteen to twenty-five. Counting various heavy artillery and miscellaneous units, the size increased finally to four thousand.[84]

Following the reorganization of the BIA as the BDA and establishment of a military academy at Mingaladon, there was a gradual increase in the size of the BDA. By March or April 1943 there were 55,000 new recruits. Stories were legion of families sending all their sons to join the BDA.

Part of the plan for the new BDA involved founding a military academy and military prep school. To foster a martial spirit among youth and give preliminary training to boys who would feed into the BDA officer class, in April 1943 a *Shonentai* or *Seifuryō* (Divine Wind Dormitory) was founded. Headed by Capt. Okada Ichirō, it was modeled after a *Yonen Gakkō* or Military Prep School in Japan. The Japanese language was taught in addition to the *Yonen Gakkō* curriculum. Recruitment was announced in Burmese newspapers. Fourteen to sixteen year-old boys were recruited through written, oral and physical examination. Thirty boys were selected from among four to five hundred applicants from all of Burma.[85]

84 Ohno Tohru, "Biruma kokugun shi", pt.2, pp.348-53.

85 Sawamoto, *Nihon de mita Biruma gun no seiritsu*, III, pp.333-4.

In September 1942 the BDA Military Academy was established at Mingaladon in the Rangoon suburbs. Candidates were selected from BDA units all over Burma. Selection was made by officers rather than through special exams. In all, five classes of officer candidates, three classes of NCOs, three classes of the *Yonen Gakkō*, and one class of air force pilots were trained in wartime Burma. Most of the members of the present Revolutionary Council and Commanders of the Burma Army are graduates of the Mingaladon Military Academy during the Japanese occupation.[86] This is true also of the Burmese diplomatic corps.

Instruction at the Military Academy in Mingaladon, unlike the original instruction at Hainan and Taiwan, was all carried out in Japanese. All officer candidates studied the language daily. Japanese Army training manuals were used, and it was anticipated in the planning of the curriculum that each class would be trained for at least one year. Candidates were selected from battalions of the BDA. The first class consisted of two to three hundred men organized in two companies. While the first class was in fact in training for nearly one year, the term of training of the second class was about half a year, and the third, fourth and fifth (and last) classes' terms were shortened successively by the exigencies of the total war situation.

From among the first class graduates were selected many of the first of two groups of thirty officers each sent to the Military Academy in Japan. Training at the Military Academy in Japan followed more closely the curriculum of the Japanese Military Academy in Tokyo, with the exception that the program for Burmans was abbreviated to one year for each group.[87]

The Military Academy at Mingaladon graduated its first class in March 1943. Thirty were selected to study in Tokyo and left for Japan on 5 June. A second group was sent later and an additional ten were sent to the Air Force Staff School in Japan.

In late 1943 and 1944 the Japanese Army had four

[86] Ohno Tohru, "Biruma kokugun shi", pt.2, pp.353-4; interview with Kawashima, July 1972.

[87] Interview with Kawashima, *op. cit.*

administrative and liaison offices through which it dealt with the BDA. All were under jurisdiction of the Burma Area Army, which came to supercede the 15th Army in actual relations with the BDA-BNA, as the 15th Army became engrossed in the Imphal offensive in 1944. First was the *Heibikyoku* (Armaments Office) staffed jointly by Japanese and Burman officers under the command of Colonel Fukui. Second was the Burma Army Advisory Department, staffed by Major Wakamatsu, Captain Noda Tsuyoshi, and Lieutenant Takahashi. Third was the *Kambu Kōhōseitai* (Officers Candidate School) at Mingaladon, under command of Major Maruoka and later Captain Kawashima. And fourth was the command of the Burma Army itself, which was similarly under jurisdiction of the Burma Area Army. After Suzuki's transfer of the command of the BIA to Aung San, the Burma Defense Army-Burma National Army, (BDA-BNA), continued under Burmese command until near the end of the war (Ba Maw was nominally Commander-in-Chief), when Aung San negotiated with Allied Commanders.[88]

After dissolution of the Baho Government in 1942 and establishment of the Military Government Bureau, the Japanese Army turned to Ba Maw for political leadership. Japanese authorities searched for him at the time of the capture of Rangoon. It was known that he had been jailed by the British. But which jail he was in was a mystery to the Japanese military. In May 1942 the *Kempeitai* discovered him in a jail at Mogok, north of Mandalay, and released him. Ba Maw was made chairman of the Preparatory Committee on Administrative Control. General Iida gave his sanction to the arrangement. At the end of May, the Thakin Party was consulted in an attempt to elicit united support for Ba Maw's leadership. Though Ba Sein and Tun Oke objected strongly, the rest acquiesced. The Preparatory Committee was accordingly ordered to establish the Administrative Council immediately following this meeting.[89]

Some residue of ill will nevertheless persisted between segments of the Thakin Party and Ba Maw, partly because

[88] *Ibid.*

[89] Izumiya, *Minami Kikan*, pp.211-12; Ba Than, *The Roots of Revolution*, pp.33-4.

the Thakins felt they were under-represented in the Ba Maw Administrative Council. When preparations were made for independence in 1943, Ba Sein and Tun Oke attempted to block Ba Maw's elevation to dictatorial control. They were consequently dropped from the list of ministers, and ultimately sent to Indonesia by the Japanese Army because of their strenuous and persistent opposition to Ba Maw.[90] Ba Sein and Tun Oke backed a scheme to restore the monarchy, which the Japanese later sought to revive in 1945.

Ba Maw's power as *Adipadi* or Head of State rested in the last analysis on the sanction of the Japanese Army. Ba Maw's old Sinyetha Party was based largely on a personal following. Professor Cady deems the party a failure "because too many Burmans distrusted the author and discounted his promises",[91] Ba Maw deployed his personal supporters through the Administrative Council throughout Burma during the Japanese occupation. To Japanese advisers, another weakness in Ba Maw's political position appeared to be the political role of his wife. The Japanese were convinced that she was handling widespread bribery in an attempt to protect her family.[92]

A further reorganization of the BDA came on the heels of Burmese "independence" of 1 August. On 15 September 1943 following Burma's independence ceremony, the first meeting of the Supreme Defense Council of Burma was held. Aung San was Defense Minister. Commander-in-Chief Colonel Shu Maung (Ne Win) announced the change of name from BDA to Burma National Army, *Bama Tatmadaw*. "The Burma National Army is resolved not to retreat but to fight through to complete victory in this war with our Japanese allies,"[93] he stressed. A month earlier Ne Win had stated to a press conference: "Burmese independence will be achieved only through victory in the Greater East Asia War".[94]

[90] Cady, *A History of Modern Burma*, p.455; Ōta Tsunezō, *Biruma ni okeru Nihon gunsei shi no kenkyū*, p.433.
[91] Cady, *A History of Modern Burma*, pp.384-5.
[92] Ōta, *Biruma ni okeru Nihon gunsei shi no kenkyū*, pp.434-5, quoting a memorandum of Takahashi Hachirō; also interview with Takahashi, 15 September 1970, Tokyo.
[93] Sawamoto, *Biruma gun no seiritsu*, III, pp.380-1.
[94] Sawamoto, *Ibid.*, p.379.

Japanese military advisors were attached to the BNA as to the BDA. Chief adviser Colonel Hayashi Masanao of BAA Headquarters was replaced by Major-General Sawamoto Rikichirō who had been adviser to the Wang Ch'ing-wei régime. Sawamoto called together Lieutenant Takahashi, formerly of the *Minami Kikan,* and Aung San and Ne Win. They discussed future plans and goals of the BNA.

Out of these talks grew a ten-page national defense plan drafted by Aung San. On 5 January 1944 Aung San showed it to Sawamoto. Sawamoto in turn drafted a proposal based on Aung San's plan which he presented to BAA commander Kawabe and his staff. It encompassed a five-year plan of expansion to four divisions and thirty-six infantry batallions. It envisioned creation of an air force, beginning with the officers trained in Japan. It contemplated use of light tanks and armored car units. It called for creation of a cavalry and training in chemical warfare, use of cannon, and enlargement of an engineering corps. It envisioned expansion of the officer corps and sending officer candidates to Japan in increasing numbers. Kawabe and the BAA staff approved Sawamoto's proposal.[95]

Sawamoto consulted too with Ba Maw, who as Chief Administrator was also supreme commander of the BNA. Sawamoto was surprised by Ba Maw's negative response to Sawamoto's explanation of the plan for the BNA. Said Ba Maw, "As the quality of the present Burmese army isn't good I don't wish to see it enlarged and strengthened." On the other hand Sawamoto heard Ne Win complain, "Ba Maw does everything British style. The only exception is his not loving the army."[96]

Ba Maw, however, depicts himself as champion of the BNA against Kawabe in 1944. It should be noted that during most of the first half of 1944 Kawabe was immersed in the staggering Imphal campaign. Ba Maw recalls a conversation with Kawabe late in 1944 after the retreat in which he, Ba Maw, criticized Kawabe for using the INA in combat but not the BDA. "You are showing your faith in the Indian Army but not in ours," complained Ba Maw to Kawabe, adding, "That's rather humiliating for us."[97] Kawabe did not change

[95] *Ibid.*, pp.385-9.
[96] *Ibid.*, p.390.
[97] Ba Maw, *Breakthrough in Burma,* p.369.

his mind, and Ba Maw attributes to his refusal to accord the BNA a real military role the turning of the BNA to politics instead. "I am convinced that the Japanese policy of creating an armed force in, for them, a totally new country and keeping it idle and exposed to all kinds of political contacts and frustrations was one of the worst blunders the Japanese committed in Burma,"[98] Ba Maw comments.

Kawabe and the BAA staff in fact had no plan to use the BNA in combat until 1945, by which time it was too late. The Japanese Army in 1944 "still did not believe the BNA was too sharp",[99] recalls one Japanese officer. The BNA was therefore used only in garrison duty, despite Sawamoto's plan for general expansion of the BNA.

Sawamoto in mid-1944 began to think of temporarily severing the BNA connection with the Japanese Army and re-training the BNA away from the front line. He considered Pyinmana or Pegu suitable sites for this plan, and broached the idea with Aung San. Actually, Japan's military plight was so critical that there was no possibility of carrying through Sawamoto's proposal.

It was Japanese procrastination in dealing with the BNA as a real military unit which contributed to the revolt of the BNA. All units trained by the Japanese were consequently politicized as well as given military training. Though they revolted against their Japanese tutors they thereby indicated that they had become a real independence army. Today Burma's army officers, members of the Revolutionary Council and diplomatic corps are nearly all from among these Japanese-trained officers. Japanese training had an effect which cannot be overemphasized: the militarization and politicization of the élite of post-war Burma.

In the chapter on revolt we will see how the BNA revolted against Japan and contacted Allied forces in early 1945. While BNA leaders thus turned back temporarily to their colonial relationship to England, the taste for independence and revolutionary momentum gained during the wartime interlude took them a full step to genuine independence in the early post-war years.

[98] *Ibid.*, p.370.
[99] Personal conversation, 21 June 1972, Tokyo.

4

Peta

Indonesia was the area of Southeast Asia richest in natural resources essential to Japan. It was also further from Japan, more populous, and had more Muslims than any other area. For these reasons Japan dealt with Indonesia differently from other parts of Southeast Asia. Indonesia was, together with Indochina, the first area in Southeast Asia where military occupation was projected in Tokyo. Still, Japan had no single cohesive "blueprint" for Indonesia, despite the use of the term by scholars of Japan's Southeast Asia venture.[1] This lack of unified policy planning was reflected, for example, in the division of occupied Indonesia into three separate administrative areas under Army and Navy command. It was also reflected in the inability of planners in Tokyo to reach any consensus on the matter of Indonesia's political status.

Formulation of Indonesia policy in Tokyo took place in a kind of vacuum. To some extent this was true of all of Southeast Asia. But for Burma and Indochina there were at least a few "experts", including civilian specialists and officers in Army and Navy General Staff Headquarters. There was, however, no "Indonesia lobby" in any official or semi-official sense.[2]

Japanese success in ousting the Dutch from Indonesia in the first few days of March 1942 made an indelible impression on the inhabitants, conditioned to regarding Dutch power as unassailable. The rapid collapse of the Dutch confirmed the nationalists' contention that the Dutch would

[1] See, for example, Thompson, Virginia, "Japan's Blueprint for Indonesia," *The Far Eastern Quarterly,* v.5, 1945-46, pp.200-7; and Benda, Harry J., *The Crescent and the Rising Sun,* pp.109, 194.

[2] See the excellent article by Nakamura Mitsuo, "General Imamura and the Early Period of Japanese Occupation", in *Indonesia,* no.10, 1970, p.4.

be unable to resist effectively. Japanese propaganda therefore elicited a favorable response initially. Japanese planes dropped posters with Japanese and Indonesian flags printed together, with the inscription, "One color, one race." When the Japanese Army advanced the Indonesian flag was hoisted. And the Djojobojo legend – a folk belief prophesying the end of white rule at the hands of yellow invaders from the north who would be the last alien rulers of Indonesia – had gained popularity since 1940.[3]

Japanese invading forces had some other advantages. At least one Japanese-trained Indonesian, Soejono, returned to Indonesia with the Japanese Army to act as liaison officer and adviser to the Military Administration in its early phase. There were also several hundred Japanese who had resided in Java for many years before the war and spoke Indonesian. Some of them, however, were lost to the occupation. Many were sent to Australia at the outbreak of the Pacific War, depriving the Japanese of a potentially valuable class of administrators. Another group of civilians being sent to Java from Tokyo was lost in the battle of the Java Sea.[4] Despite these losses, however, Japan enjoyed a propaganda advantage in Indonesia when the invading forces arrived. The rapid eviction of the Dutch and the Japanese slogan "Asia for Asians" had raised the hopes of the populace and heightened nationalist aspirations. A further significant stimulus to the nationalist movement was the release from prison after several years in exile of prominent leaders Sukarno and Hatta.

The nationalist movement was of course a major source of potential support for Japanese occupation policies. The facts of the development of the Indonesian nationalist movement are well known but can be reiterated briefly here. Focussed in its early stages on Muslim groups and prophesies of the coming of a Messiah, the beginnings of Indonesian nationalism were heavily religious and mythic. Sarekat Islam was the organizational center of popular political aspirations in the second decade of the twentieth century. Some leaders were

[3] Benda, Harry J., "The Beginnings of the Japanese Occupation of Java", *The Journal of Asian Studies,* v.15, 1955-56, p.545.
[4] *Ibid.*, p.543.

already perceived by the Dutch as being opposed to the government and accordingly were expelled from Java. Marxist theory was introduced to students in Java at the same time, and in 1920 the Communist Party of Indonesia, KPI, was founded.

Tjokroaminoto, charismatic leader of the Sarekat Islam, numbered among his students Sukarno who had been raised by his father to serve the fatherland. Declaring that the Sarekat Islam had the same goal as the KPI, the welfare of the Indonesian people, Sukarno nevertheless chose the path of the Sarekat Islam rather than the KPI. Graduating from technical college with an engineering degree in 1926, Sukarno was by then already highly politicized and believed both in Marxist progress and in Allah.

In 1927 Sukarno founded what was to become the National Party of Indonesia, PNI, favoring non-cooperation with the Dutch colonial power. Though the party was regarded as virtually illegal it nevertheless burgeoned. Because of Sukarno's activities he was imprisoned, and while he was in prison a split occurred within the nationalist movement, in part precipitated by a critical letter from Mohammed Hatta in Holland. By the time of Sukarno's release in late 1931 popular adulation of him had reached the proportions of a cult.

The other branch of the nationalist movement, led by Sjahrir, emphasized the class struggle by the exploited masses as the only struggle for freedom that had a real chance of success. Sjahrir and Hatta, both Sumatrans, were far more Europeanized than the Javan Sukarno, whose base of power was the Indonesian masses and their mythic traditions. Sukarno sought to radicalize and stimulate the masses to a readiness for struggle.

For Sukarno and much of the nationalist movement Islam became increasingly important in the decade of the thirties. Sukarno began to view Islam as a source of added political support, and even of progress. Sent into exile in Benkulen, Sumatra, Sukarno was by the time of this outbreak of the Pacific War a major force and leading nationalist with whom the Japanese had to reckon. Sukarno combined the potency of nationalism, Islam and Marxism.[5]

Nationalist leaders were brought into relationship with Military Administration in an advisory capacity, though some, like Sjahrir, continued to operate underground as well. An order of 24 September 1942 directed that a research institute be established to preserve the peace of Java as a supply base and to ensure the use of civilian leaders. The Research Institute, headed by Hayashi Kyūjirō, was staffed with nationalist leaders Hatta, Sukarno, Sutardjo and Mansur. The stated goals of the Institute were to do research on old customs and also to facilitate transmission of views of the populace to upper levels in the administration.

Nationalist hopes engendered by the Japanese arrival were almost immediately dashed by orders issued two weeks after the surrender of Batavia (Djakarta). One was a Japanese decree forbidding display of the Indonesian flag. The same day another decree banned "any kind of discussion or organization, suggestions or propaganda concerning the political organization or administration of the country". Controls over communications media were also established. Bernhard Dahm suggests that the reason for repression of political activity was not because of a "well-prepared occupation policy", as Benda suggests, but for the opposite reason – the absence of a cohesive policy necessitated a moratorium on political activity.[6]

The evidence for Dahm's interpretation is clearer, in view of the general lack of Japanese policies for dealing with Southeast Asia. Japanese journalists had contracts with Indonesian nationalists before the war, but once hostilities erupted they were sent to Australia or became inactive. Some of those in Australia later participated in Military Administration in Indonesia, but none was consulted during pre-war policy making for the Netherlands Indies.[7]

There were, of course, individuals who served as military attachés in Batavia in the thirties. These included Army

[5] See analysis of Sukarno by Dahm, Bernhard, *Sukarno and the Struggle for Indonesia Independence*, Ithaca: Cornell University Press, 1969.

[6] *Ibid.*, pp.220-1. In the view of this author Dahm's evaluation was closer to the truth, posited as it was on the absence of a clear policy.

[7] Nakamura, "General Imamura", p.4.

Lieutenant-Colonel Nakayama Yasuto and Navy Captain Maeda Tadashi. While neither played any political role in Tokyo early in the occupation, Maeda had significant contacts with nationalists later in the occupation. Still, there were no "political officers" in Tokyo who were personally committed to Indonesia immediately prior to the war. Besides this combination of lack of expertise, information, or commitment to Indonesia there was the significance of the Dutch East Indies as a source of critical petroleum. For all these reasons Japanese policy toward Indonesia's political status never achieved a clear consensus in the early months of the war or later.

There is no doubt that Japan's major goal in the Netherlands East Indies was acquisition of natural resources to strengthen Japan's war potential and support her industrial establishment.[8] An ancillary objective was attaining autarchy within the Greater East Asia Co-Prosperity Sphere.[9] Japan conceived of Java in particular as a supply base for the rest of Southeast Asian military operations, and of Sumatra as a primary source of petroleum.

Japan's economic interest in Indonesia was reflected in several policy decisions made in Tokyo and steps to implement them prior to the outbreak of war. In early 1940 the government despatched two economic missions to the East Indies, the first headed by Kobayashi Ichizō, Minister of Commerce and Industry in the Konoye cabinet, and the second by Yoshizawa Kenkichi, former Minister of Foreign Affairs.[10] In a 27 July 1940 decision of the Liaison Conference it was indicated that Japan would take diplomatic steps to win the resources of the Dutch East Indies and other parts of Southeast Asia and would resort to military force if necessary.[11] The ABCD encirclement which cut Japan off from the East Indies created a crisis in Japan which

[8] Benda, Harry, James Irikura and Kōichi Kishi, *Japanese Military Administration in Indonesia: Selected Documents,* pp.12-13.

[9] *Ibid.,* p.17.

[10] Van Mook, H.V., *The Netherlands East Indies and Japan, Their Relations, 1940-1941,* pp.24, 64.

[11] Waseda University Social Sciences Research Institute, comp., *Studies in Japanese Military Administration in Indonesia,* translated by the U.S. Department of Commerce, p.105.

helped precipitate the decision for war.

On the eve of war the basic principles of military administration determined jointly by the Army and Navy on 26 November 1941 were three: restoration of public order, rapid acquisition of defense resources, and self-supporting operations in the field.[12] But problems of how to interpret and execute these basic goals were legion and were in fact never resolved until the end of the war.

One example of the lack of unity in policy was the administration of General Imamura Hitoshi, first commander of the 16th Army, which occupied Java. Imamura was of the relatively enlightened persuasion, prompted also by the "Guidelines for Occupied Areas" (*Senryō Tōchi Yōkō*) adopted in Tokyo, that attention should be paid to the customs and traditions of inhabitants if occupation were to succeed in its aims. He was sympathetic with Indonesian goals, including the aspiration for independence. His attitude had romantic and emotional overtones. He reports soon after occupation of Java being approached by a venerable village official, who related to Imamura the Indonesian folk legend of Djojobojo. Imamura reportedly replied to the old man through an interpreter, "The ancestors of the Japanese people went to Japan from this island by ship. You and the Japanese are brothers. We fought the Dutch in order to regain your freedom."[13]

Because of Imamura's sympathies he earned a reputation for his "mild policy" in Java. He encountered opposition from younger staff subordinates immediately. They felt that for the 16th Army to make its authority felt among Indonesians, the Army should first adopt harsh, oppressive measures, which could later be alleviated if necessary. Imamura consistently opposed this line of argument, whether proposed by his own staff, by the Southern Army, or by Imperial General Headquarters in Tokyo. He held that his policy was based on the "Guidelines for Occupied Areas" decided in Tokyo, and that unless those guidelines were

[12] *Ibid.*, p.106.

[13] *Imamura Hitoshi Taishō kaisōroku,* v.10 (Recollections of General Imamura Hitoshi), p.118; see also Imamura Hitoshi, *Tatakai o owaru* (The struggle ends).

revised there was no possibility of choosing an alternative policy.[14] Furthermore, he was convinced his policy was correct.

Though Imamura's word as commander carried in Java, there were repercussions shortly from Tokyo. The Army sent three politically powerful advisers to Java, with instructions to meet with Imamura to discuss his policy. The three envoys who arrived in Java in mid-April were Kodama Hideo, ex-Minister of Internal Affairs, Hayashi Kyūjirō, former Ambassador to Brazil and economic envoy to Java, and Kitajima Kenjirō, Deputy Minister of Colonial Affairs. They immediately took Imamura to task, reporting criticism of his policy from Tokyo, Saigon and Singapore on the grounds that the Japanese Army was failing to demonstrate its authority in Java. Imamura in his own defense pointed out that there was no comparison between the situation in Java, where the inhabitants had cooperated with the invading Japanese forces, and in Singapore, where the Chinese had resisted. Furthermore, in terms of access to Java's natural resources, it was not feasible to be more oppressive.[15]

This visitation was followed by a barrage of top staff officers from Tokyo. First came Chief of General Staff Sugiyama, with Colonels Hattori and Takeda. Sugiyama reported there would be yet another delegation from Tokyo – General Muto Akira, Chief of Military Affairs, and General Tominaga of Personnel in the Army Ministry. Imamura replied that he would only change his policy on orders from Tokyo, and in that eventuality he would request a transfer. General Muto later told Imamura that the situation had changed since the adoption of the original guidelines in Tokyo, and that he could no longer base his policy on them. Muto insisted that the Japanese Army had succeeded with such unexpected rapidity that it was not necessary to pay attention to the situation of the inhabitants. Imamura remained adamant, arguing that if oppressive means were used Japan would have another insoluble "China Incident" on her hands. The debate remained deadlocked.[16]

[14] *Ibid.*, pp.154-5.
[15] *Ibid.*, pp.155-8.
[16] *Imamura Hitoshi Taishō kaisōroku,* v.10, pp.158-62.

Despite Muto's hard line with Imamura, on his return to Tokyo Muto defended the Java policy on grounds that it was too early to judge results and it was unreasonable to expect the same policy to work in every area of the south. When Muto was sent to Sumatra he dispatched a letter of apology to Imamura, adding that he was modelling his policy in Sumatra after Imamura's. General Terauchi from Southern Army Headquarters visited Java in June and was also satisfied with what he saw. Imamura was commended by Colonel Ishii Akiho in the Southern Army Staff for being the only commander in the South to follow the guidelines. Imamura felt vindicated.[17] There is some debate about whether Imamura's later transfer to command of the 8th Area Army was in some way a demotion, but Imamura denies this, pointing out that the 8th Army was a larger force than the 16th.

The relative liberalism and idealism of Imamura was still apparent within upper echelons of the 16th Army after he was transferred from command in Java in November 1942. A document titled "Suggestions on the Future Status of Java" appears with the sub-title "Osamu Group Commander" and was possibly written before February 1943.[18] It advocates a declaration that self-government shall be granted to Indonesia in the near future because: 1) Indonesians are friendly and expectant toward Japan; 2) there is danger of discontent over material problems; 3) it is necessary to win the hearts of the inhabitants to further development of resources; 4) Japanese are over half the population of occupied Southeast Asia and their wishes cannot be ignored; 5) Japan should not make the same mistakes the Dutch made of viewing Indonesians as ignorant; and 6) the declaration should be made now rather than in the face of British and American counter-attacks, when it will be too late.[19] This liberalism, however, was not shared elsewhere generally in the Army or Navy, in 1942 or even in 1944.

In fact widespread dispute raged within the military

[17] *Ibid.*, pp.166-7.

[18] So thinks Nakamura: "General Imamura", p.22.

[19] Benda, Irikura, Kishi, *Japanese Military Administration in Indonesia*, pp.237-9.

establishment concerning the current and future jurisdiction over the East Indies. The initial goals of the Navy Ministry, as detailed in the "Outline on the Conduct of Military Administration in Occupied Areas" of 14 March 1942 stated that in areas occupied by the Navy (which included the Celebes, Moluccas, lesser Sunda Islands and Borneo), authority would be directed toward permanent possession under Japan's control and integration into the Japanese Empire.[20] While the Navy expected its territory to be incorporated into the Empire with Japanese victory, the Army was more equivocal. Some, for example General Muto in Sumatra and General Yamashita in Malaya, assumed that Sumatra and Malaya would become Japanese territory. Others were not so decisive and thought the status of Sumatra, for example, should be settled after the war.[21] The general tendency was to equivocate and avoid making hard-and-fast commitments for as long as possible. The major reasons for this temporizing were that sharp conflicts of opinion between the Army and Navy, and within the Army between levels of command, could not be resolved.

By 1943 the original goals of military administration receded into the background as defense and supply requirements assumed greater importance. When Japan accorded "political participation" to Indonesia, Malaya and Indochina, the reservation was made that military administration would be continued "for the time being".[22] And a Liaison Conference in January 1943 decided that "areas of strategic importance which must be secured by the Empire for the defense of Greater East Asia shall be incorporated into the Empire." The Army and Navy thus both sanctioned the principle of incorporation of certain areas into the Empire. Though no areas were specified in this decision, petroleum resources made the East Indies an area of prime strategic significance for Japan. And the Navy was already on record in 1942 specifically in favor of permanent possession of areas

[20] *Ibid.*, pp.26-9.

[21] See discussion in Kanahele, George, *The Japanese Occupation of Indonesia*, pp.39-40.

[22] Waseda, *Studies in Japanese Military Administration in Indonesia*, trans., p.123.

it occupied.

In January 1943 Tōjō announced in the Diet that Burma and the Philippines would be given independence within the year. This was a great shock to Sukarno, Hatta, and other leaders who had cooperated with Japan in the faith that Indonesia would be the first nation in Southeast Asia to attain independence. Indonesian confidence in the Japanese was greater than in other areas of Southeast Asia, and the disappointment was correspondingly heightened. Sukarno and Hatta sought the intervention of Miyoshi Shunkichirō, a political adviser with the Military Administration who had many years' pre-war experience in Japanese diplomatic service in Java. They demanded to know the reason for such an insult to Indonesia, and asked that further radio broadcasts of the announcement be stopped. Miyoshi took them to see Colonel Nakayama, head of General Affairs and Major-General Kokubu, to appeal the Indonesian case. The staff officers acknowledged Japan's indebtedness to Indonesian cooperation, but pointed to the difference in the intensity of war in Burma compared to the relative calm of Indonesia following the occupation.[23] Miyoshi could only offer sympathy.

The standard Japanese reply to Indonesian pleas for concrete steps toward independence was that the first step was to win the war. Unless victory were first achieved there would be no genuine independence. This response had succeeded between the beginning of the occupation and January 1943 in encouraging among Indonesians the faith that they would be the first nation in Southeast Asia to become independent.

The reason for not allowing independence, as outlined by the Foreign Ministry a year later, was "based simply on the obvious necessity of meeting the various military and economic requirements of the Empire in the successful prosecution of the Greater East Asia War".[24] And in General

[23] Miyoshi Shunkichirō, "Jawa senryō gunsei kaikoroku", (Recollections of military government in occupied Java), *Kokusai mondai* (International problems), pt.4, Aug. 1965, no.65, pp.65-6.

[24] Benda, Irikura, Kishi, *Japanese Military Administration in Indonesia*, p.240.

Staff Headquarters there was also an understanding that the "Greater Indonesia movement should be curbed as much as possible."[25] The word "Indonesia" was accordingly avoided in Japanese pronouncements.

Nationalist leaders continued to appeal to Military Administration authorities through various channels. A new avenue of appeal was opened by the May 1943 visit of Greater East Asia Minister Aoki Kazuo to Java. Hatta, Mansur and Dewantro appealed to Aoki in a concerted effort to elicit some response from Japan, as a logical sequel to Japan's liberation of Indonesia from Holland. Hatta particularly requested permission to use the national flag and anthem to raise public morale. He also pleaded for Indonesian administrative unity. Hatta warned Aoki that unless Japan took concrete measures, continued cooperation of the Indonesian people could not be guaranteed. Aoki could only promise to put the Indonesian case strongly before authorities in Tokyo.[26]

Aoki's visit was followed a month later by Tōjō's statement in the 82nd Diet on 16 June 1943 that the Javanese people would be given a chance for "political participation". The announcement regarding Burma and the Philippines, increasing Javanese demands for independence, and continuing differences within the military establishment over policy for Indonesia, all had added to pressure for Japan to make some concession in form if not in substance. Tōjō in his address before the Diet outlined Japanese plans for various areas of occupied Southeast Asia. For Sumatra, Java, Borneo and the Celebes (he avoided the term "Indonesia") he announced that Japan would move towards "political participation of natives during this year in conformity with the desire of the natives". Further, "where Java is concerned, it is anticipated that this will be realized as promptly as possible in view of its cultural level and in response to the confidence

[25] Quoted from an August 6, 1942 decision, "Outline for Leadership of the Peoples of Greater East Asia", by Nakamura, in "General Imamura", p.5.

[26] Miyoshi, "Jawa senryō gunsei kaikoroku", pt.4, Aug. 1965, no.65, pp.67-8.

shown by the people".[27] The particulars, however, were not spelled out, and 16th Army Commander Harada Kumakichi simply conveyed this general message to some ten leaders.

Tōjō's announcement was followed by a personal inspection tour of the Southern Theater in July. His itinerary included a visit to Java, where he spoke before a welcoming rally attended by fifty thousand people. Tōjō expressed thanks for past cooperation and stated that concrete steps toward political participation would be implemented in the near future. He failed to mention any details of "concrete steps", and this failure again deflated Indonesian hopes. Harada on 1 August defined political participation to mean establishment of a General Council in Djakarta, state councils, and a municipal council in Djakarta as advisory organs for Military Administration. Formation of these councils was implemented two months later.[28]

In 1944 some of the differences within the Army between levels of command regarding independence for Indonesia came into focus. This came about because of requests by the field forces, and because the Army recognized that it would be difficult to elicit continued cooperation without making some concrete commitment regarding independence. Other factors were economic deprivation and political malaise, which heightened following the announcement regarding Burma and the Philippines. Army Civil Administrator Hayashi Kyūjirō on 20 March 1944 proposed a "Plan for the Governance of Java" to alleviate discontent. He suggested a declaration that Java (not Indonesia) be granted independence "when their preparatory education for the post-war future has been completed". The independent nation, according to Hayashi's plan, would consist of Java, Madura, and Bali.[29] This proposal of Hayashi was reflected in a document of 5 August 1944 emanating from the Supreme War Council, which suggested granting independence in three stages: first to Java, Madura and Bali,

[27] Benda, Irikura, Kishi, *Japanese Military Administration in Indonesia*, p.51.

[28] Miyoshi, "Jawa senryō gunsei kaikoroku", pt.9, pp.67-70.

[29] Benda, Irikura, Kishi, *Japanese Military Administration in Indonesia*, pp.245-6.

second to the Celebes and North Borneo, and finally to all areas of the Netherlands East Indies.[30]

In August 1944 the Army finally came around to support of independence, though it opposed Hayashi's suggestion of making only Java independent. The Navy still opposed any form of independence. The Foreign Ministry was strong in its insistence on independence. Disagreements between the Army Ministry and the more moderate Greater East Asia Ministry were conciliated during discussions between 29 August and 2 September. They feared a demand for a unified state and still avoided the term "Indonesia".[31]

The 2 September agreement embodied the final decision prior to a public pronouncement by Premier Koiso in September. The Navy, nevertheless, still refused to capitulate, but attached reservations to the paragraph in the 2 September agreement which stipulated "the areas to be given independence shall be the former Dutch Indies (except New Guinea)".[32] The Navy also had reservations on preparations for independence in areas other than Java, and in fact clung to the hope of permanent possession of areas under Navy jurisdiction.

Premier Koiso announced in the 85th Diet in early September that Indonesia would be granted independence. The announcement at face value was electrifying to Indonesians. He did not, however, specify either the date for independence or the form of the new independent state. And the minutes of the Supreme War Council indicated that the matter was far from settled. The minutes stated, "As a result of the statement the question as to the areas and time to grant independence shall continue to be studied, and it has been agreed that a statement will be announced in the Diet."[33] Koiso was therefore purposely vague on these points.

It was not possible to clarify these questions, since the deadlock between the Army and Navy continued over details

[30] Waseda, *Indonesia ni okeru Nihon gunsei shi no kenkyū* (original), p.370.
[31] Waseda, *Japanese Military Administration in Indonesia*, trans., pp. 376-8.
[32] Waseda, *Japanese Military Administration in Indonesia*, trans., p.379.
[33] *Ibid.*, p.370.

of the manner and timing of independence. The deadlock continued until March 1945 (and even beyond) when the establishment of the Independence Preparatory Investigation Committee was finally announced. Meanwhile, Military Administration instead stressed the importance of the political participation system, in an attempt to capture the national consciousness and divert attention from the question of independence. The 25th Army in Sumatra and the Navy were opposed even to offering political participation, to say nothing of independence.[34]

The Foreign Ministry, following the Koiso announcement, advised that recognition of independence would have several advantages: it would win popularity, it would prove the constancy of Greater East Asia principles, and it would contribute to the unity of Greater East Asia. Disadvantages, on the other hand, would be loss of freedom in military and economic measures, and the possibility that it might be construed only as a tactical measure.[35]

Differences between the Army and Navy, between the service ministries on the one hand and the Foreign and Greater East Asia Ministries on the other, between the 16th Army in Java and the 25th Army in Sumatra, and between levels of command, including the Southern Army Headquarters in Saigon and 7th Area Army Headquarters in Singapore, could not be resolved. These differences were exacerbated by the deterioration in the economy and war situation generally in 1944 and even late 1943. When the problem of independence was taken up at the policy-making level in Tokyo, the emergence of a consensus took over a year because of these discussions. Disagreement prevailed between the Army and Navy Ministries in Tokyo, between the two forces in the field, and between the Southern Expeditionary Fleet and the Naval Attaché's Office in Djakarta.

In Java differences between Army and Navy authorities regarding attitudes toward independence were further aggravated by the sympathetic attitude of Navy Laision officers. Admiral Maeda in a letter to Navy Military Affairs Office at the time of the visit of Greater East Asia Minister Aoki

[34] *Ibid.*, p.372.
[35] *Ibid.*, p.374.

complained of the attitude of the Japanese Military Administration toward Indonesia's future, and further requested that the use of the national flag and anthem be allowed. This led to further Army – Navy confrontation in the field.[36] This sympathetic attitude of the Navy Liaison Office in Java had virtually no influence on the Army. Maeda's position was furthermore basically at variance with Navy Headquarters in Tokyo. His influence was greater, in fact, with nationalist leadership than within the Japanese military establishment.

These dissensions were not simply an example of idealism versus bureaucratism, though this disparity added another dimension to the conflict. The continuation of high-level discussions on the problem meant that it also became intertwined with carrying out military operations and with the political-economic situation. Perpetuation of the discussion also heightened the gap between the aspirations of Indonesian nationalists and the response of Japan.[37]

While Indonesian nationalists again took heart at the Koiso declaration, it was yet another six months before any further hints of independence were made by Japanese authorities, and even then in such a way as to succeed only in further aggravating nationalist opinion.

The difficulty was that except for propaganda pronouncements by Tōjō and Koiso, Indonesian independence was handled in Tokyo basically as a problem of strategy. This was true near the end of the war as the Supreme War Council in Tokyo foresaw an Allied invasion. There was still a desire to preserve first Java then all of Indonesia as a campaign supply base. This was reflected in the discussions of August 1944. This engendered disparities between the Foreign Ministry and the service ministries, which were acting from operational concerns. Thus the Foreign Ministry proposal of 28 August mentions that the Army Ministry had not yet completed its study of the matter, and that the Navy Ministry opposed changes in the *status quo*. Though the Greater East Asia Ministry was a source of pressure for independence, even that

[36] Waseda, *Indonesia ni okeru Nihon gunsei shi no kenkyū,* p.400.

[37] See discussion in Waseda, *Japanese Military Administration in Indonesia,* pp.408-9.

ministry's proposal of 25 August stipulates that the date of independence should not be announced by the government. The Army and Foreign Ministries concurred with this suggestion.[38] As the war situation deteriorated further in 1945 there were nearly daily meetings in Tokyo and Djakarta on the subject of Indonesian independence.[39]

When on 10 March 1945 the announcement was made of a new policy regarding independence, it in fact served as a further irritant to nationalist opinion. It was announced that a triple policy would be implemented: 1) establishment of a Preparatory Committee for Independence; 2) establishment of a School for the Founding of the Nation; and 3) relaxation in restrictions on speech. A Preparatory Committee of exclusively Javanese members was in fact appointed. But Tokyo still made no specific decisions regarding the critical matters of the timing or scope of independence. It was therefore apparent that this announcement was yet another palliative designed to placate nationalist opinion without taking any practical steps.[40]

Thus until the end of the war no solution was reached to the problem of disjunction in goals of the Indonesian nationalists and the Japanese military authorities. None of the many pronouncements made by Japanese Military Administration ever satisfied nationalist aspirations for independence. Politically the Japanese and Indonesian leadership remained fundamentally deadlocked for the duration of the war. Within the framework of this deadlock, however, Japan endeavored to mobilize Indonesian opinion via mass organizations and to organize and train a number of military and para-military units. In this area the legacy of Japanese military occupation of Indonesia is still being felt. The "Three A" Movement, Java *Hōkōkai, Miai, Masjumi,* and *Putera* were all Japanese-fostered efforts at mass political mobilization, with varying degrees of success. (Bernhard Dahm, Benedict Anderson, George Kanahele, and Harry Benda have all discussed aspects of these organizations. See bibliography.)

Because of the ambiguities regarding Japanese policy

[38] Benda, Irikura, Kishi, *Japanese Military Administration in Indonesia,* p.249.

[39] Waseda, *Indonesia ni okeru Nihon gunsei shi no kenkyū,* pp.370-9.

[40] Waseda, *Japanese Military Administration in Indonesia,* pp.412-4.

towards Indonesian independence, none of the pronouncements emanating from Tokyo had a great impact on nationalist aspirations, except as a negative stimulus. It was rather at the local level, in Java, that Japan provided the greatest positive stimulus to the nationalist goal of independence. This incentive was provided through the organization and training of a volunteer army, the *Sukarela Tentara Pembela Tanah Air,* or Army Defenders of the Homeland, abbreviated as *Peta.* The creation of this army was the work of an intelligence agency, the *Sambōbu Tokubetsu-ham,* or *Beppan* in abbreviation.

Tokumu kikan were used by the Army throughout Southeast Asia. But since the term *tokumu kikan* had acquired pejorative connotations in China, the Japanese Army in Java was careful to avoid the term. Under the intelligence office of the 16th Army Headquarters, it was instead variously called *Isamu Bunshitsu* or *Isamu Gunshitsu* in the earliest stage.[41] In April 1942 the organization was put under command of Colonel Murakami of intelligence, and the name *Beppan* came into use. Like its counterparts elsewhere in Southeast Asia, the *Beppan* was staffed with graduates of the *Nakano Gakkō.* Attached to the *Beppan* were interpreters from civilian life, many of whom had lived in the Netherlands before the war or had studied the Malay language at the Tokyo or Osaka University of Foreign Languages. *Beppan* headquarters was at Bandung.

The *Isamu Bunshitsu* in Bandung in early 1942 was staffed with two officers, two soldiers, six interpreters, two Japanese women married to Dutchmen, and one tall Dutchman known as "Nishimura", who had studied in Osaka and had a Japanese wife.

One of the two Japanese officers in Bandung in early 1942 was Lieutenant Yanagawa Munenari. Yanagawa had already reconnoitered roads and collected weapons disguised as an Indonesian. He was a graduate of the *Nakano Gakkō.* He had also assembled a group of six young Indonesians and

[41] Yanagawa Munenari, "Jawa no Beppan", (The *Beppan* of Java), *Shūkan Yomiuri, Nihon no himitsu sen* (Weekly Yomiuri, Japan's secret war) 1956 special issue, p.153; and interview with Yanagawa Munenari, 17 January 1971, Djakarta.

taught them judo, *sumo, kendō,* and some language in the evening. He used these young men for gathering intelligence. Other early *Beppan* activities included collecting Indians for the Indian Independence League, which was active in Malaya and Thailand with the *F Kikan.* Major Fujiwara came to Java, and Yanagawa discussed with him the subject of independence for Southeast Asia. Yanagawa dates his concern with independence to this meeting with Fujiwara.[42]

With the reorganization of the *Beppan* in April 1942 its head office was moved to Djakarta and four branches were established in Bandung, Semarang, Jogjakarta, and Surabaja. Throughout the war the *Beppan* functioned as an intelligence agency. In late 1943 and 1944 it was primarily involved in training *Peta.* In April 1945 it underwent another change of name and was known as *Nanseitai.* It functioned increasingly as a defense organization after late 1944.[43]

The staff of the *Beppan* included, besides Lieutenant Yanagawa (later Captain), Captain Marusaki (later Major) recruited locally, Lieutenants Tsuchiya, Yonemura, and Miyake, and 2nd Lieutenants Hoshizume, Yoshitake and several interpreters. One of the interpreters was Ono Nobuharu (Abdul Hamid Ono), a Japanese Muslim who had lived in Indonesia for many years.

Among its special activities the *Beppan* was assigned to make advance preparations for an offensive against Australia. In this connection Yanagawa and others were ordered to produce a film, "Calling Australia", and Yanagawa spent some time in Tokyo in 1944 on this assignment.

Officers on the *Beppan* staff were assigned to train the officer corps of the Java Defense Volunteer Army (Java *Bōei Giyūgun,* as the Japanese called it), or *Peta.* Because of Japanese reluctance regarding Indonesian independence, there was no direct invlovement of *Beppan* members with the independence movement, unlike the training of the INA and BIA by *tokumu kikan* staff in Burma. Creation of *Peta*

[42] *Ibid.*
[43] Interrogation of Captain Tsuchiya Kisou April 3-8, 1947 by Dutch authorities: Nishijima papers, Waseda University Social Science Research Institute.

was nevertheless an indirect encouragement to the war of independence.

In 1945 the *Beppan* had four sub-sections. One section under Lieutenant Tsuchiya was in charge of counter-intelligence, and Tsuchiya was also briefly attached to the *Peta* units training in Bali. The *Nami Kikan* (Wave Agency) under Lieutenant Yonemura was in charge of collecting intelligence on enemy operations along the coast and of preventing infiltration of enemy agents along the shore.[44] Lieutenant Yanagawa was in charge of the training and development of *Peta.* Finally, 2nd Lieutenant Yoshitake was in charge of overseas Chinese operations, including organization of a small Chinese guerrilla force.[45] At this time there were 150 staff members in *Beppan,* with fifty additional special duty members, including some Indonesians.

Yanagawa opposed giving military training to Chinese youths. There were, however, three Chinese Youth Training Centers in Djakarta, Semarang and Tjirebon. About five hundred Chinese were trained in each program, as at the Tanggerang Youth Training Center (see below, p.103). From late 1944 until the end of the war they were given some guerrilla training, though, like *Hizbullah,* they were trained not with weapons but with bamboo spears.[46]

The *Sendenbu,* or Propaganda Section, was a separate organization from *Beppan* and was a section of the Military Administration. It conducted ideological warfare and consisted of five sub-sections: General Affairs, News, Films, Propaganda, and Broadcasting. In April 1943 the broadcasting section became an independent bureau.[47]

The *Sendenbu* was headed by Shimizu Hitoshi, somewhat notorious for having been a pacification officer with the Japanese Army in north China. He was also affiliated with the Imperial Rule Assistance Association and with the *Tōa Remmei* (East Asia Federation). Shimizu remarked, "I have had affiliations with the *Tōa Remmei* in the past, and I have

[44] Interrogation statement of Captain Yonemura Masao, I.C. 0064600-006464 in Nishijima papers, Waseda University.

[45] Waseda, *Japanese Military Administration in Indonesia,* p.470.

[46] Interview with Tsuchiya Kisou, 29 March 1972, Mishima City, Japan.

[47] Waseda, *Indonesia ni okeru Nihon gunsei shi no kenkyū,* pp.247-8.

had an idea of developing in Indonesia a spiritual movement of popular enlightenment which may be termed an Asian movement."[48]

Japanese forces feared a possible united Indonesian leadership, or even several groups of Indonesian leaders who might turn against Japan.[49] Shimizu and the Propaganda Section therefore turned their attention to the formation of mass organizations which would mobilize political support for the Japanese occupation. Among these were the "Three A" Movement, *Putera,* the Java *Hōkōkai,* and the *Shishintai.* Mass rallies, parades, and meetings were the techniques used in these movements. Even the Japanese realized, however, that the first of these, the "Three A" Movement, was a failure, and nationalist leaders continually criticized the movement. When the Propaganda Section conceded the failure of the movement, it was replaced by *Putera* on 9 March 1943. *Putera* also failed, for Japan had no clear idea of the future of Java and still refused to allow nationalists the use of national anthem and flag.[50]

Shimizu's more nefarious activities included organizing several secret societies. Most significant of these was the Black Fan, to prepare Indonesians for a war of independence under the guidance of Japan. His staff also organized the *Miai,* whose goal was uniting all the people and which automatically included all *Kiais* (Muslim religious functionaries), though there was no attempt at secrecy in the *Miai.* When *Kiais* attempted to act politically on their own, the *Miai* was transformed in November 1943 into the *Masjumi.* Another group, the Black Snake, was directed toward Indo-Europeans, and had its headquarters in Bogor. Perhaps most notorious of all Shimizu's groups was the Chin Pan, a Chinese secret society.[51]

The Navy was less active than the Army in using

[48] Waseda, *Japanese Military Administration in Indonesia,* pp.471, 334.

[49] *Ibid.,* p.333; Benda, "The Beginnings of the Japanese Occupation of Java", p.544.

[50] Kahin, George Mc.T., *Nationalism and Revolution in Indonesia,* p.103; Waseda, *Japanese Military Administration in Indonesia,* p.39, pp.344-54.

[51] Interrogation of Shimizu Hitoshi, November 27, 1945, I.C. 0065-00591, Nishijima papers, Waseda University.

intelligence agencies in Southeast Asia, though it is not true that the Navy took no interest in such activities. The Second Southern Expeditionary Fleet in Indonesia established in Djakarta a Naval Attaché's Office for liaison with the 16th Army, for negotiating with the Army for requisitioning supplies to be shipped to Navy-controlled areas. Headed by Rear-Admiral Maeda Takashi, it was charged also with intelligence activities and particularly with investigating political and economic conditions. The Attaché's office included three departments: External Affairs, Economy, and Investigation, staffed mostly by civilians. Indonesians attached to the Investigation Section of this office included prominent leaders such as Wikana and Subjardjo, who played important roles in the independence movement.

The most signfiicant role of the Attaché's Office in connection with the independence movement was the establishment of the controversial *Asrama Merdeka Indonesia,* or School for a Free Indonesia. Founded in December 1944, the school was under the supervision of Yoshizumi Tomegorō, a journalist in pre-war Java, and Nishijima Shigetada, with the Indonesian staff headed by the well-known Communist, Wikana. Lecturers included Subardjo, Wikana, Sukarno, Hatta, and Sjahrir.[52]

Admiral Maeda was especially sympathetic with nationalist aspirations and gained the confidence of nationalist leaders to a degree that enabled him to play a unique role individually when independence was finally proclaimed by Indonesians at the end of the war.

Maeda had close ties with Dr. Achman Subardjo, a Dutch-enducated Indonesian with romantic memories of a honeymoon in Tokyo when he was a correspondent there in the mid-thirties. In the Navy's Research Bureau (Subardjo's term) with Hatta, he was assigned to investigate the consequences of regulations of the Military Administration, particularly in the economic sphere. The Navy was interested both in access to supplies and in understanding the nationalist movement. It was Subardjo who selected the faculty and students of the *Asrama Merdeka Indonesia.*[53]

[52] Waseda, *Japanese Military Administration in Indonesia,* p.473; and interview with Dr. Subardjo, 22 January 1971, Djakarta.

[53] Interview with Dr. Subardjo.

Admiral Maeda hoped in the Navy-sponsored school to inculcate a spirit of independence within the context of Asian solidarity. The Indonesian faculty were remarkably free to teach what they chose, and what they did teach and the motivations of Maeda have been the subject of much scholarly debate, partly because of Wikana's connections with the PKI. Kahin alleges that Maeda and the intelligence officers who assisted him emphasized the study of communism in the curriculum in order to infiltrate the PKI underground and to align it with Soviet policy and turn it against Britain and America in anticipation of their invasion of Indonesia. Kahin suggests that Maeda foresaw a Japanese-Soviet alliance after the war.[54] Such a theory of motivation is tenuous in view of the general orientation of the Navy, and has in fact been refuted by both Maeda and Subardjo.[55] Maeda states, "I avoided moulding this training after the Japanese pattern, neither had I the intention of employing the youths for some definite purpose after finishing the course. The idea that I should make use of the Communist Party for the furtherance of my schemes shows unexpected bias against me. Nobody who knows me will countenance the idea for a movement."[56]

Apart from the propaganda effort, Japan trained in Indonesia several armed and para-military bodies which had even more lasting impact. The deterioration of the military situation with the fall of Guadalcanal, followed by the loss of more ships and personnel at Attu and Rabaul, led to the transfer of personnel from Java, including the 16th Army Commander, Major-General Imamura Hitoshi. General Imamura and others were diverted to the 8th Area Army. The shortage of armed strength on Java was increasingly apparent to staff officers of the Southern Army in late 1943. This lack of defensive strength in Java prompted several

[54] Kahin, *Nationalism and Revolution in Indonesia*, pp.115-20.

[55] Dr. Subardjo has refuted Professor Kahin in a series of newspapers articles and in an interview with this author: 22 January 1971. Maeda also denied these motives in a conversation with Yoneda, Press Attaché of the Japanese Embassy in Djakarta: author's interview with Yoneda in Djakarta, January 1971.

[56] Interrogation of Vice-Admiral Maeda Takashi, 9 November 1946, I.C. 011352-011356, Nishijima papers, Waseda University.

proposals in Tokyo, Saigon and Java to supplement deficient Japanese troop strength. Japan's motivation in the creation of *Peta* was thus more military than political, unlike the INA and BIA.

The first units of young Indonesians trained to supplement deficient defensive power were the *heiho,* or auxiliary troops. During the war an estimated 25,000 Javanese were trained and attached to the Japanese Army for various kinds of supplementary and special services. These included air defense units, tank units, transport units, roadbuilding in Burma and Thailand (where they functioned more as *rōmusha,* or forced labor) and even for counter-intelligence in the *Kempeitai.*[57]

The *heiho* system was adopted in Indonesia, Malaya and Burma in the spring of 1943. In Indonesia *heiho* troops were selected through recommendation by Education Office heads of Military Administration in the various states. Those selected were bachelors aged sixteen to twenty-five who had over half a year's education in Japanese. They were given two months' military training, after which they were deployed in units attached to the Japanese Army. Indonesians, however, came to view the *heiho* as simply laborers in uniform; the prestige value of the system from the Indonesian viewpoint was negative, though their value for the Japanese as labor was high.[58] The *heiho* system was of military and economic value to the Japanese and was used widely throughout Southeast Asia. Thousands of *heiho* sent to Burma and Thailand from Indonesia never returned after the war. Many of them were presumed lost on the notorious Thai-Burma Railway.

While the *heiho* and *Peta* were trained out of military necessity, there were also numerous para-military groups organized and given some military training by the Japanese. These included 5–600,000 unarmed *Seinendan* (youth groups), 1,286,813 *Keibōdan* (Civilian Defense Corps), 80,000 *Shishintai* (Pioneer Corps), 50,000 *Jibakutai* (Suicide Corps), 50,000 *Hizbullah* (Muslim Youth Corps), and 50,000 *Gakutai* (Student Corps). It should be noted, however, that

[57] Waseda, *Indonesia ni okeru Nihon gunsei shi no kenkyū*, p.191.
[58] Pauker, Guy, "The Role of the Military in Indonesia": Waseda, *Japanese Military Administration in Indonesia,* p.94.

these units were armed and trained only with bamboo spears and that the training was at best perfunctory. The goal of training these groups was mutual cooperation and support for the war rather than military participation as fighting units.[59]

As a result of directives from Tokyo and Saigon, on 3 October 1943 16th Army Commander Lieutenant-General Harada Kumakichi announced the formation of a volunteer force "based on the spirit of common defense of Greater East Asia and in response to the intense desire of fifty million Indonesians for defense of the homeland" (no mention was made of independence). Japanese military advisers would be attached to the units, which would be under command of the 16th Army.[60] Many complex ingredients went into the making of this decision.

Sukarno in his autobiography claims credit for single-handedly selecting a nationalist leader, Gatot Mangkupradja, to take the initiative in proposing a volunteer force. Sukarno recalls having been requested by the Japanese High Command to help in finding the proper officer candidates. Says Sukarno, "I handpicked leaders like Gatot Mangkupradja, the PNI rebel with whom I'd been arrested in 29. He was made head of *Peta.* I looked for young men whom I could control and who could eventually become the heroes of our Revolution. I singlehandedly proposed the future colonels and generals of our Republican Army back in the fall of 1943."[61]

A contradictory version of who exercised the initiative in the creation of *Peta* is told by Gatot Mangkupradja. Gatot was a nationalist who had been educated in European-style schools in Batavia and Bandung. He joined the PNI in 1927 at the time of its establishment and became secretary of the executive section of the party. In 1929 he was arrested with Sukarno and imprisoned for three years. After release from prison he continued nationalist activities until 1933 when he

[59] 16th Army Headquarters, "All Kinds of Armed Bodies", Nishijima papers, Waseda University.

[60] Waseda, *Japanese Military Administration in Indonesia,* pp.194-5.

[61] *Sukarno, an Autobiography as told to Cindy Adams,* pp.186-7. Actually, Gatot had virtually nothing to do with *Peta* once it was created, belying Sukarno's statement that he picked Gatot as its commander.

opened a pharmacy in Bandung and left the front ranks of nationalist leadership. In this year he also visited Japan as a member of a commercial delegation.[62] He felt Japan had a deep understanding of Asian nationalism and in Tokyo he met with other Asian nationalists such as Rash Behari Bose. From the Japanese wartime viewpoint Gatot had the triple advantage of credibility with nationalists, sympathy toward Japan, and not being so powerful a leader as to be uncontrollable.

Gatot's first action was spontaneous and without Japanese prompting. Gatot was opposed to a proposal for introducing conscription in Java, a suggestion made in the newspapers by Raden Sutardjo Kartohadikusumo. Gatot found the proposal contrary to the views of the Indonesian Nationalist Party and to those of Sukarno, who had declared conscription beneficial for an independent nation but not for a subject people.[63]

Still a different version is told by Togashi Takeomi, a Japanese shopkeeper in Tjiandjur and a friend of Gatot. Togashi later became an interpreter for *Beppan.* Togashi believes it was he who first convinced Gatot that an army is a requisite for every independent nation, in a conversation prior to Gatot's proposal. Togashi reports being asked by Gatot for suggestions as to how he should go about making the proposal for an army.[64]

Tsuchiya Kisou, another member of *Beppan* who was involved in training programs at Tanggerang, Bogor, and Bali recalls that *Beppan* selected Gatot and that the initiative for and substance of Gatot's proposals were provided by *Beppan.* But, adds Tsuchiya, *Beppan* tried to create the impression that the initiative was Indonesian, not Japanese.[65]

[62] Kurasawa Aiko, "*Peta* and the 1945 Revolution", pt.1 (in Japanese), M.A. thesis, 1971, Tokyo University. Mrs. Shiraishi (née Kurasawa) Aiko is the most knowledgeable authority regarding the history of *Peta.*

[63] Gatot Mangkupradja, "The *Peta* and My Relations with the Japanese; a Correction of Sukarno's Autobiography", in *Indonesia,* no.5, 1968, p.115.

[64] Interview with Togashi 15 March 1972, Tokyo.

[65] Tsuchiya related that *Beppan* attempted to elicit a similar proposal through the Sultan of Jogjakarta, but that this approach was not fruitful; interview with Tsuchiya 11 March 1972, Mishima City.

Gatot made his proposal for a volunteer force through the newspaper *Tjahaja*, whereupon he was interrogated by the *Kempeitai*. He was released and the following day was asked to go to the *Beppan* office in Djakarta, where he met Lieutenant Yanagawa.

Once in the *Beppan* office there is no question that the initiative was Japanese. Major-General Satō asked Gatot to put his proposal in writing and send it to Army Headquarters and to the Military Administration. Gatot states that in order to demonstrate his sincerity he pricked his arm and wrote the petition in his own blood, a procedure which impressed his Japanese witnesses.[66] The petition was printed in *Indonesia Raja* in Djakarta and other parts of Java. Similar petitions appeared in newspapers all over the country

Gatot states that he was on friendly terms with Yanagawa and discussed with him the hope for independence, and that Yanagawa supported the idea.[67] Yanagawa, on the other hand, states that it was not feasible even to mention independence, to say nothing of discussing it with Indonesians. Independence was contrary to Army policy. Further, states Yanagawa, he took a warning from the case of Colonel Suzuki Keiji in Burma, who when he tried to redeem his promise to the BIA of independence, lost his position as head of the *Minami Kikan*.[68] Yanagawa was nevertheless impressed with the possible analogy between his own role and Lawrence of Arabia (an analogy made also by Fujiwara and Suzuki of their own roles) and would no doubt have liked to appear in this light to his Indonesian friends, as he intimated in a conversation with Gatot. Since it was impossible to encourage Indonesians directly in their aspirations for independence, Yanagawa and his advisers decided to promote Islam as the spiritual basis for the volunteer army.[69] Actually, it was a matter of higher Japanese policy to

[66] Gatot, "The *Peta* and My Relations with the Japanese", pp.115-29; a lawyer, Supankaat, advised Gatot to specify that the volunteer army be developed into a national army later.

[67] *Ibid.*, p.118.

[68] Yanagawa Munenari, *Rikugun chōhōin Yanagawa Chūi* Lt. Yanagawa, an Army intelligence officer), p.112.

[69] *Ibid.*, pp.113-14.

promote Islam in Indonesia, contrary to former Dutch policy.[70] In any case, Yanagawa and others were aware of the nationalist aspiration for independence, and it may be that Gatot was asked to put his request in writing to avoid conveying the impression that the Japanese Army was simply creating a puppet army.

One author has it that the Japanese kept early plans for creation of *Peta* secret from first-rank nationalist leaders such as Hatta and Sukarno, and selected a leader of medium rank in Gatot.[71] Actually, however, only a few days elapsed between the time of the conversation between Yanagawa and Gatot and the publication of the Harada announcement and call for volunteers in the press. Thus it became public knowledge almost immediately, and Sukarno and Hatta certainly read of it if they were not in on the initial conversations.[72]

The idea that Military Administration tried to keep plans for formation of *Peta* secret from top rank nationalist leadership is also contradicted from another source. Tanaka Masaaki cites a conversation between Military Administrator General Yamamoto Moichirō and three nationalist leaders, including Sukarno. According to Tanaka, the discussion centered around organization of a volunteer army. Sukarno favored a genuine national army in name and actuality, and suggested that only a reconnaisance unit be attached to the Japanese Army.[73]

Several factors, then, conditioned the Japanese decision to organize and train a volunteer army in Java. The major concern was the inadequacy of Japanese forces for the defense of Java, with only 10,000 troops stationed there. Another increasingly important concern was the role of Java as a logistical and resource supply base for all of Southeast

[70] See discussion on Japan's policy toward Islam in Benda, Harry J., *The Crescent and the Rising Sun.*

[71] Kanahele, George, *The Japanese Occupation of Indonesia,* p.120; see also Kurasawa, "*Peta* and the 1945 Revolution", pt.1.

[72] Yanagawa also reports being ordered to help plan the increase of defensive strength on Java but to avoid stimulating consciousness of independence, use of the flag, or liaison with nationalist leaders such as Sukarno and Hatta: Yanagawa, *Rikugun chōhōin,* pp.110-11.

[73] Tanaka Masaaki, *Fusetsu yonjūnen no yume; hikari mata kaeru* (A forty-five year dream: the light returns again), p.101.

Asia. Still another factor was the dissatisfaction and malaise among nationalists, particularly after the January 1943 announcement of independence for Burma and the Philippines. These factors combined in the thinking of several individuals in Tokyo, Saigon and Java to produce the decision to form and train *Peta.*

In Tokyo and in Saigon forces were set in motion by the tour of inspection by Lieutenant-General Inada Masazumi, Deputy Chief of Staff of the Southern Army, in the summer of 1943. Inada's thinking was influenced by the inadequacy of Japanese troop strength on Java. On 13 June Inada visited anti-aircraft units on Java, some of whom were *heiho.* He recognized that this effort was inadequate to deal with the problem. Inada felt this was symbolic of the whole defensive situation in Southeast Asia. He also feared the small Japanese force in Java would not even be able to preserve peace and order. He was convinced that *heiho* units could not solve the problem and might even be dangerous or useless. There would have to be trained units officered by Japanese.[74] Inada discussed the problem with Harada during his tour in Java and with General Tanabe of the 25th Army in Sumatra. Harada was enthusiastic. He saw the plan as an effective means of guiding Indonesian opinion, particularly if Japan did not grant independence. It would therefore facilitate his own duties as commander of the 16th Army. Inada impressed on both Harada and Tanabe the importance of eliciting the trust of the inhabitants if Japanese control of Southeast Asia were to be successful.[75] Inada also broached the subject with Tōjō and General Satō Kenryō, Chief of Military Affairs in the Army Ministry, both of whom were touring Southeast Asia. Inada suggested formation of indigenous forces in Java, Sumatra, Malaya and North Borneo. Tōjō gave his approval but said Japan could not furnish weapons for such armies.[76]

[74] Inada Masazumi, *Shōnan nikki* (Singapore diary) unpublished, v.1, pp.245-55; v.2, pp.380-410.

[75] *Ibid.*

[76] Interview with General Inada 24 November 1970, Tokyo. Kanahele states that Inada had to face opposition of members of his own staff on the grounds that it was dangerous to arm indigenous people, and that Tōjō also refused to supply funds: Kanahele, *op. cit.*, p.119. Inada denied that Tōjō had refused funds, in an interview with this author.

None of the three generals appears to have considered the precedents of the INA or BIA during this discussion. This is curious since these armies were both trained and functioning two years prior to Inada's proposal. This is added evidence of the lack of an overall Japanese plan for Southeast Asia, in the matter of independence armies as in other areas.

Another proposal made in Tokyo was that of Colonel Nishiura Susumu at the end of 1942, also aimed at supplementing Japanese troop strength in Southeast Asia. Inada, however, did not mention this Tokyo proposal as a factor in his thinking. Actually, Colonel Nishiura's suggestion influenced not the organization of volunteer armies, but rather the formation of *heiho* units.[77] Furthermore, IGHQ had already in June 1942 issued an order to the Southern Army for the formation of "indigenous armed bodies" (see Chapter One).

Initiative in the creation of *Peta,* then, resulted from a combination of pressures: from Tokyo, from Southern Army Headquarters, through General Inada, from 16th Army Commander Harada, from *Beppan* officers like Lieutenant Yanagawa, and from nationalists in petitions like Gatot's.

A few days after General Harada's 3 October announcement, a call for volunteers went out. Qualifications for officers were irrespective of education, but they should be of firm resolve and robust health, of no special age. For non-commissioned officers they were to be under twenty-five, bachelors, and in good health. In the selection process there was special emphasis on recruiting volunteers who had a strong national consciousness. This was determined in oral interviews by Japanese recruiting officers. Recruitment was carried out in every state. Graduates of the Tanggerang Youth Training Center formed the nucleus of the officer corps. It was Japanese policy to avoid inducting candidates who had been trained by the Dutch in the *KNIL* or those with Dutch-style education, on the premise that their awareness as Indonesians would be underdeveloped, and that they would be less likely to be anti-Dutch.[78] There were no

[77] This conclusion was corroborated in an interview with Colonels Fuwa and Fukushige of the Bōeichō Senshishitsu (Defense Agency, War History Library) 7 February 1972, Tokyo.

[78] Yanagawa, *Rikugun chōhōin,* p.122; Waseda, *Indonesia ni okeru Nihon gunsei shi no kenkyū,* p.194.

Peta battalions raised in Eastern Indonesia, the Northern Celebes, or Amboina, as these were Christian areas and traditional recruiting grounds for the Dutch *KNIL*.

Youth training centers all over Java fed into *Peta* officer units. In the youth training centers hundreds of boys were given para-military training and indoctrination, to enable them to assist in civil defense. The initial training period was six months. Following this period of training some boys were selected from each center to be sent to the *Tanggerang Seinen Dōjō* in early 1943 near Djakarta. From the Jogjakarta Youth Training Center, for example, some outstanding individuals were selected to go to Tanggerang. Among them were Zulkifli Lubis and Kemal Idris, both of whom became high-ranking army officers in Indonesia after the war.

The entrance exam for the Tanggerang Center consisted of a physical examination and a number of oral questions regarding Dutch colonialism. The Japanese throughout Southeast Asia, in civilian and military training programs alike, attempted to ascertain the spirit, motivation, and anti-colonial sentiment of the Southeast Asian recruits. They did this through apparently perceptive oral interviews, judging by the subsequent career success of graduates of the training programs.

The Tanggerang Center opened in January, 1943, several months prior to the decision to train a volunteer army. Fifty boys between the ages of sixteen and twenty-two were selected from centers all over Java for military training by officers of the *Beppan.* Lieutenants Marusaki and Yanagawa were in charge of the Tanggerang Center. Boys were brought to Tanggerang in such secrecy that in many cases their parents were unaware of their destination. For six months they were put on a spartan regimen which began at 6:00 a.m. and included elementary military techniques, hard physical exercise and instruction in counter-intelligence. Despite the harsh discipline, all but two of the first group successfully completed the course. Graduates included Lubis, Idris, Daan Magot, and Suprajada Jonosewojo, all of whom rose in the ranks of the post-independence army. Lubis recalls that after the recruits passed the first course they did practical training for one month, then re-entered the Center as the senior class, where they remained until October or November. The program was so successful that the 16th Army and *Beppan* staff decided to train a second class of thirty-five recruits.[79]

Yanagawa and other instructors used "special methods" designed to inculcate self-confidence in the trainees. The boys were also taught *sumo* wrestling. Yanagawa reports that staff officers from the 16th Army were impressed with the visible changes in the physique of the trainees. Yanagawa also saw that the boys were given some training in intelligence, military science, espionage, and the Japanese language. Because of Japanese policy regarding independence it was not possible to conduct any real political indoctrination of the trainees.[80] The success of the Tanggerang Center program in turn encouraged General Harada in the plan to create a volunteer army, and the Tanggerang graduates in fact fed into the officer candidate group for *Peta* training. When the decision to organize *Peta* was announced in October, the Tanggerang Center was closed and Yanagawa and others were transferred to Bogor to open the Officer Training Center there with star pupils from Tanggerang.

The Bogor *Renseitai* (Officer Training Unit) was opened at the end of 1943 with Japanese instructors like Yanagawa from the *Nakano Gakkō*. Officer candidates were of three ranks: *daidanchō,* or commanders of battalions of approximately five hundred men; *chūdanchō,* commanders of companies of one hundred and fifty to two hundred men; and *shodanchō,* commanders of smaller platoons. Site of the training program was an old *KNIL* army barracks. In all some seventy Indonesians were trained as *daidanchō,* two hundred as *chūdanchō,* and six hundred and twenty as *shodanchō,* plus two thousand non-commissioned officers, *bundanchō.*[81] Seniors from the graduating class at Tanggerang, including Lubis, became assistant instructors in the

[79] Interview with Colonel Zulkifli Lubis 20 January 1971, Djakarta; Kanahele, *op. cit.,* p.118; Yanagawa, "Jawa no Beppan", p.154.

[80] Interview with Yanagawa 17 January 1971, Djakarta.

[81] Interview with Effendy Pandjipurnama in Bandung, 30 January 1971; see also Pauker, Guy, "The Military in Indonesia," in Johnson, John, *The Role of the Military in Underdeveloped Countries,* pp.190-5.

shodanchō training program.

Daidanchō candidates were leaders of some stature in their communities, men of middle age. Commander of the Djakarta battalion, for example, was Kasman Singodimedjo, a law professor, later Attorney General, and a well-known Muslim leader who had been jailed by the Dutch. Training of the *daidanchō* was brief, about one and a half months. Their value was primarily in adding their prestige in the recruitment of large numbers of men for their local *daidan*.

Chūdanchō were trained at Bogor for nearly three months. They learned enough Japanese to understand command words. The training was tough but effective, according to those who went through the program. A day's training lasted all day, with only six hours for sleep. Long all-day marches were followed by running, in turn followed by jumping into a cold pool. The program was not elaborate technically and did not include study of military strategy and science. It was designed to teach Indonesians to endure hardship with a spirit of sacrifice and self-reliance. The emphasis on spirit, *seishin* (*semangat* in Indonesian), was universally noted by all interviewees trained here as in other Japanese programs in Southeast Asia, whether military or civilian. Training was designed to produce physical courage and stamina as well.

From the group of *chūdanchō* came high-ranking generals of the Indonesian post-war army. Among them were General Bambang Sugeng, later Chief of Staff and subsequently Ambassador to Brazil, the Holy See, and Japan. Another *chūdanchō* was Lieutenant-General P.H. Djatikusumo, son of the Susugunan of Solo. He was exceptional in that he had studied in Holland for four years and had had reserve officer training under the Dutch in Bandung. Generally those with Dutch education or training were avoided in selection of officer candidates. Another notable exception was Suharto, later President of Indonesia. Suharto graduated from the Bogor Renseitai as a *shodanchō,* and later in September 1944 was specially selected for *chūdanchō* training, despite his earlier Dutch service.[82] It was impossible for the *Beppan* to

[82] Tsuchiya Kisou, *Indoneshiya giyūgun to Suharto Daitōryō* (The Indonesia volunteer army and President Suharto), Tokyo: Indonesia Center, n.d., pp.8-9.

avoid completely those with Dutch training and experience, and if a candidate sufficiently impressed *Beppan* interviewers he was recruited. The goal of *shodancho* training was to give them as nearly as possible the equivalent of the technical training of their Japanese counterparts.

General Djatikusumo was thus in a unique position of being able to compare Dutch and Japanese training. He observed, "The Dutch were better in theory, but the Japanese were more practical. We didn't learn from the Dutch how to make an army. What we learned from the Japanese was more important: how to create an army from scratch and lead it. We learned how to fight at company level, how to recruit soldiers, and how to devote yourself to your country."[83]

The *chūdanchō* were selected through local government officers and influential relatives. They were born between 1912 and 1918, which meant they were in their thirties during the war years. After three months' training they were sent out to recruit and train their own *chūdan.* They had no training manuals, but had to rely on the knowledge acquired during their own training at Bogor. Weapons were also different from those used in Bogor training.

Shodanchō were generally under twenty and were recruited from throughout Java, through officials and relatives. A second class in the Bogor *Renseitai* was for *shodanchō* rank only. There were also noncoms, *bundanchō,* trained at Bogor.

There was no training of a general or central staff of officers for *Peta* until close to the end of the war. Nor was there direct communication between or among *daidan.* Japanese advisers were attached to all *daidan,* and all official liaison and communication among *daidan* was carried out by Japanese officers. There is some disagreement as to whether the Japanese intention was not to train a general staff, in order to avoid potential danger.[84] In fact, a core group of

[83] Interview with General Djatikusumo, 19 January 1971, Djakarta.

[84] Kanahele believes this was possible: *op. cit.,* p.126; Kasman Singodimedjo feels it was fear that prompted the Japanese failure to train a general staff. The suggestion has also been made that it was simply lack of time for adequate training that was responsible for lack of more advanced training for staff officers: interview with Colonel Lubis, January 1971.

officers was trained beyond the level of the Bogor program for *chūdanchō.* This group, including Lubis, some contend, was intended to form the nucleus of a general staff.[85] Part of this group with advanced training, including Lubis, were sent to Bali in 1944 as special instructors for the three *daidan* organized on Bali.

Officer training at Bogor began on 18 October and ended 7 December 1943 with the commissioning of officers in the presence of General Harada.

In February 1944 Lieutenant Tsuchiya went with Japanese assistants and four Indonesians, graduates of Tanggerang and Bogor, to train three battalions on Bali. Lieutenant Tsuchiya graduated from the *Nakano Gakkō* in June 1940 and was assigned to Taiwan where he studied Indonesian and worked on advance preparations for training the Burmese "thirty comrades". In mid-November 1941 he was ordered to the 16th Army in Java, but his arrival was delayed until after the Singapore campaign. The Indonesians who accompanied Tsuchiya to Bali included Lubis, Idris, and Magot, and they trained some fifteen hundred men in three *daidan* on Bali.[86]

The officer class training for the Bali *daidan* was completed in June and the recruitment and training of *daidan* began. There were some problems distinctive to the Bali *daidan* which were not encountered in Java. For one thing, Army authorities in Bali were not overly enthusiastic about the *Peta* project there. The five hundred Army troops stationed in Bali were not under 16th Army command but under the 2nd Division. Bali itself was under Navy jurisdiction. However, Tsuchiya recalls that the Navy commander was very cooperative in providing facilities for training the Bali *daidan.* Another factor was that the religion of Bali was Hindu rather than Muslim, and when officers from Java were given pork they were horrified. Yet another problem in Bali was the lack of transportation and communication facilities. Recruits arrived on foot in the absence of roads.

Though Bali was under Navy jurisdiction, the training of the Bali *daidan* was carried out under the aegis of the

[85] Kanahele, *op. cit.*, p.126.
[86] Kanahele, *op. cit.*, p.130; interview with Colonel Lubis, January 1971 Djakarta, and with Tsuchiya 11 March 1972, Mishima City.

Beppan, as with the Java *daidan,* in a striking case of Army-Navy cooperation. In other areas under Navy control there was no attempt to organize or train indigenous volunteer units. The Navy did not feel it had the personnel, resources or time to expend in such an effort or that the results would justify it. The Navy appears also to have distrusted Christian Ambonese and Menadonese who would have provided prospective recruits in Navy areas.[87]

Training given the Bali *daidan* lasted six months. The men were trained to work on coastal defenses, simple fortifications made of tree trunks, rocks and dirt. This reflected the basic purpose of creating *Peta,* namely coastal defense.

A total of thirty-five *daidan* were trained initially in Java. In August 1944 with the training of additional officers, another twenty *daidan* were created. Finally, in November of the same year a third group of officers trained eleven *daidan,* bringing the total to sixty-six, besides the three *daidan* formed on Bali.[88] This made the size of *Peta* in Java at full strength 33,000, plus the 1500 on Bali.

The career of Zulkifli Lubis is illustrative of all stages in the training and development of *Peta.* Born in 1923 in Sumatra of parents who were both teachers, he moved to Jogjakarta for high school. It was there that the outbreak of war in the Pacific caught him. When Japanese forces entered Java all schools were closed for several months. Unlike other students who returned to their homes during this period, Lubis remained in Java hoping for a chance to study. He was rewarded when he was admitted in the last half of 1942 to the Jogjakarta *Seinen Dōjō* for six months of para-military training which included "imprinting of war-mindedness on peoples' minds", military training, and training for security and defense. After graduation from this course Lubis was selected via examination for further training, and entered the first class of the Tanggerang *Seinen Dōjō.* After the first class Lubis was again among those selected for advanced training, while most of the graduates dispersed to their home states and their own homes. He then entered the Bogor Officers,

[87] Kanahele, *op. cit.,* p.130.

[88] 16th Army Headquarters, "All Kinds of Armed Bodies"; and Pauker, "The Military in Indonesia", p.190.

School, where he participated in instruction of the *shodanchō*. There also he was among those selected for advanced training. He then entered a small Research Group (*Kenkyūhan*). From the Research Group a few, again including Lubis, were sent to train *daidan* officers on Bali. A second sub-group of the Research Group remained in Bogor. Part of the second class at Bogor and the second sub-group of the Research Group were given nine months' guerrilla warfare training as instructors in the First Task Force in 1945.[89]

Lubis in post-war Indonesia served as Deputy Chief of Staff and was a leader of the Outer Islands Revolt in 1956, for which he was imprisoned by Sukarno. At the time of the interview he headed a special intelligence unit under Suharto.

During 1944 a training unit began to train a small group of staff officers. The *Kambu Kyōikutai* (Officer Training Unit) was organized in January 1944 at Bogor. This program was under the command of Captain Yamazaki. Yanagawa was during this period in Tokyo making propaganda films.

During 1945 the 16th Army took further emergency measures to train two additional military and para-military units as a precaution against anticipated invasion by Allied forces. A plan was drafted in December 1944 to train guerrilla units for the defense of all Java. Called the *I-go Kimmutai,* or First Task Force, the program was headquartered at Bandung, with branches at Salatiga and Malang. Captain Yanagawa again was in charge of Bandung training and nine or ten Japanese officers and non-coms were attached to each center. Each center trained fifty candidates, selected from the Bogor *Peta* officer group, and later two hundred more were added to each center. Unlike *Peta,* the First Task Force was under direct command of the 16th Army. The first training period was January through June 1945, and training was more extensive than for *Peta* officer candidates. Following the first training period officers were to be deployed to their home areas on 15 June, where they were to recruit and train men in specialized guerrilla techniques such as infiltration, liaison, communications and disguise. No uniforms were worn by this unit. Training was focussed on surprise attack techniques.

[89] Interview with Colonel Lubis, January 1971.

Guerrilla training was a military technique introduced and used by the Japanese throughout Southeast Asia, especially in the last stages of the war. Despite the natural suitability of much of island and continental Southeast Asian terrain to guerrilla warfare, it had not been widely used by the Dutch or British colonial armies. This was one factor which accounted for the rapid victory of Japanese invading forces over the Dutch Army and Dutch-trained *KNIL.* Guerrilla training was therefore one significant legacy of Japanese training acknowledged by *Peta* officers.[90]

A second plan for improving eleventh-hour defense capability of Java in 1945 was linked with the Japanese policy of promoting Islam. This was a plan to organize an island-wide Muslim youth group para-military training program. The plan was actually devised late in 1944 and took shape partly through conversations between Captain Yanagawa and the Religious Affairs Section of Military Administration. The emphasis was to be on physical training, with some elementary military training to raise the defensive capability of inhabitants generally. Recruitment of boys between sixteen and twenty-five began in February 1945 through recommendations of Muslim schools and teachers. Training was held in Bogor, training ground of the *Peta* officer corps. Known as *Hizbullah,* actual training was conducted by *shodanchō* from the First Task Force. Each of twelve *shodanchō* took charge of fifty trainees. The daily routine was similar to *Peta* training, except that greater emphasis was placed on Muslim prayers and customs. Afternoons were devoted to military training using bamboo spears and wooden guns rather than actual weapons. Special instruction was given in making Molotov cocktails, explosives and, bombs and in blasting.[91]

Yanagawa forbade the *shodanchō* to strike trainees, a practice common in *Peta* training as in the Japanese Army, but which violated Muslim precepts and sensibilities. When one Javanese *shodanchō* suggested to Yanagawa that striking trainees might help to inculcate a more martial spirit,

[90] Interviews with Generals Bambang Sugeng, Djatikusumo, and Hidajat in Djakarta, January 1971, and with Soeparjadi, former *shodanchō,* and Effendy Pandjipurnama in Bandung, 30 January 1971.

[91] Yanagawa, *Rikugun chōhōin,* pp.171-2.

Yanagawa instead recommended *sumo*. Yanagawa was impressed with results of the first month's training of *Hizbullah* youth. After three months the *Hizbullah* members were returned to their homes, where they in turn organized units. *Hizbullah* grew to fifty thousand from the original nucleus of trainees of twelve *shodanchō* from the First Task Force.[92]

Through the sympathetic inclinations of men like Miyoshi, Imamura, and Maeda, Indonesian nationalism received some positive stimulus. Beyond this impetus the creation and training of *Peta* and several para-military groups gave Indonesian aspirations revolutionary capability. The training of several thousand Indonesians by the Japanese turned out to be a potentially explosive development, given the political climate of Java in 1945. *Peta* in particular discovered that it had the military techniques to make revolution a reality. Japanese troops were the first to confront this revolutionary potential, which was later directed at the returning Dutch.

[92] *Ibid.*, pp.173-4. Gatot claims credit for having suggested to the Japanese not only the idea for formation of *Peta* but also in late 1944 for the creation of an armed force of Muslim youth. He recalls that Yanagawa agreed to the proposal and promised to talk it over with the general: Gatot, "*Peta* and My Relations with the Japanese", pp.126-7.

5

Volunteer Armies in Malaya, Sumatra, Indochina, Borneo and the Philippines

In Malaya, Sumatra, and Indochina Japanese motives in training volunteer armies, *giyūgun,* were basically military. In Malaya and Sumatra nationalist movements were not as developed or united as in Java, Burma, or among Indians in Southeast Asia. This called for different policies from those Japan devised in Burma, Java, or for overseas Indians. In Malaya, for example, communal divisions between Malay, Chinese and Indian made combined political action or a united nationalist movement against British colonial rule out of the question.

Ibrahim bin Yaacob was the leading Malay nationalist. Born in Pahang in 1911, he became a teacher and journalist, writing for Malayan newspapers *Majelis* and *Warta Malaya.* He opposed colonialism and was anti-British in particular. He was closely connected in his beliefs and activities with Ishak bin Haji Mohammed, whose anti-British political novels were banned by the government. Journalists in Malayan nationalism played a role comparable to lawyers in nationalist movements in other parts of Southeast Asia.[1]

In May 1937 Ibrahim and Ishak led other pro-Indonesian nationalists in founding the *KMM (Kesatuan Melayu Muda*), or Malay Youth League, in Kuala Lumpur. The *KMM* was the most active organization among Malays. Other leaders of the group, men like A. Karim Rasjid, were immigrant Indonesians. The goals of the *KMM* were independence for Malaya and union with Indonesia. Articles in the party organ, *Warta Malaya,* were strongly anti-British.[2]

All leaders of the *KMM,* including Ibrahim, were arrested

[1] Soenarno, Radin, "Malay Nationalism, 1900-1945", *Journal of Southeast Asian History*, vol.1, no.1, 1960; pp.1-27.
[2] *Ibid.*, p.22.

and imprisoned prior to the eruption of war in the Pacific. The *KMM* was nevertheless small and never captured the majority allegiance of even the Malay peasantry. The British planned to send Ibrahim to prison in India, but due to the rapid collapse of British power in Malaya, the plan could not be executed. With the fall of Singapore all *KMM* leaders were released and resumed their political activities. Some joined the Malay Communist Party, while others worked with the Japanese occupation. Some also cooperated with the Japanese as a cover for underground activities. Japanese military administration, deciding the *KMM* represented a potential danger, banned it in June 1942.

Chinese and Indian organizations were for the most part separate from Malay groups and from each other. Japanese policy throughout the war was suspicious of Chinese in Malaya and Singapore, as elsewhere in Southeast Asia, and Chinese were notably absent from many Japanese-sponsored civil and military groups. The Chinese responded to Japanese occupation policies with various types of resistance. One of the most notable was the Malay People's Anti-Japanese Army, formed in the jungles by the Malay Communist Party, with the assistance of British-sponsored Force 136. Estimates of the size of the MPAJA vary from four thousand to seven thousand.[3]

Given the divisive communal situation and absence of a united, nationalist movement in Malaya, Japanese plans for propaganda warfare and intelligence in both Malaya and Sumatra – administered together under the 25th Army – were much less elaborate than for Java, Burma, or the overseas Indians. Nor was there any Japanese intent to recognize Malayan independence until near the end of the war, in July 1945. This difference was symbolized in the use of the term *giyūgun,* volunteer army, rather than *dōkuritsugun,* independence army, the term for the BIA. *Giyūgun* organized in Sumatra, Malaya, Indochina and Borneo were all created to supplement deficient troop strength for coastal defense in anticipation of renewed Allied counter-attack.

[3] See discussion in Tregonning, K.G., *A History of Modern Malaya.* Tregonning's estimate of the size of the MPAJA is 7,000 (p.282); Chin Kee Onn puts it as 4,000 in *Malaya Upside Down,* p.121.

None of them was conceived as battle-ready fighting units; they were rather designed to perform auxiliary functions and supplementary coastal defense services. Other evidence of this difference is that almost no graduates of the *Nakano Gakkō* were involved in training the Sumatra and Malaya *Giyūgun,* since political warfare goals were only incidental. Further, only one Malay graduated from the Military Academy in Japan by contrast with seventeen Indonesians, sixty Burmans, thirty-five Indians, and fifty-three Filipinos.

With Thailand, Japan in December 1941 concluded an agreement which allowed Japanese troops to pass through Thailand into Malaya, following landings on the coast of Thailand. Japan's *quid pro quo* was to offer to transfer to Thailand two Thai-speaking Burmese Shan States, Kengtung and Mongpan, together with four Malay states. The two Shan states were in fact administered separately from the other Shan states during the war.

Since Japan did not establish a military administration proper in Thailand, neither was there any attempt to recruit and train an independence army there. Thailand was, after all, the only independent nation in pre-war Southeast Asia. It would not have been politically feasible for Japan to have altered this fundamental status. Nor was there any military necessity to do so, so long as Japan was able to get permission for the passage and stationing of troops.

As it was, many leading Thais were angry at Japan because they felt they had been conquered. Though they could not express their sentiments openly, the Free Thai Movement and Force 136 flourished underground.

In Malaya and Sumatra, plans for *giyūgun* were conceived in late 1943, pursuant to the year-old order from IGHQ and more immediately precipitated by General Inada's proposal to organize indigenous armies in Southeast Asia. The formation of *Peta* and of the *Giyūgun* in Malaya and Sumatra therefore had the official imprimatur of both IGHQ and Southern Army Staff Headquarters by September 1943. The Southern Army Staff, through the direction of Colonel Kaizaki, prepared an "Outline for the Organization of Native

Armies" in September 1943.[4]

Following Tōjō's approval of Inada's plan, Inada visited Southeast Asia again in August 1943. He discussed the plan briefly with General Tanabe Moritaka, 25th Army commander in Bukittinggi in Sumatra. He also met General Muto Akira, commander of the Imperial Guards Division at Medan, and reported Muto was in agreement with the plan also, raising no objections. On 22 August, Inada visited Borneo Defense Army Headquarters at Kuching, where he talked with commanding General Yamawaki Masataka and Chief of Staff General Managi Takanobu. On 23 August, he went on to Djakarta, where he consulted with General Harada Kumakichi and Chief of Staff General Kokubu Shinshichirō. On 26 August, Inada had similar conversations with Military Administration officers in Malaya.[5] In January 1945, Inada revisited all these Army commands to see how the training programs were progressing, as well as to discuss logistic problems for sustaining the war.

In Malaya, two types of volunteer unit were organized and trained, the *Giyūgun* and the *Giyūtai* (volunteer corps). The *Giyūtai* was conceived as a corps for the defense of the coastline and for the preservation of public peace and order. It was a nation-wide unit, organized at the city, town and village level. There was no central training camp, and training thus differed from one unit to the next. The training period was sporadic, in some areas once a week, in other areas for continuous periods of a week or two. Boys were selected from each family in designated villages. Since the training was not for regular, extended periods of time the recruits continued to live at home. The weapons used in training were small arms captured from British forces, and hunting guns.

[4] None of the documents from Southern Army Staff Headquarters regarding creation of indigenous armies is extant, according to Colonel Imaoka Yutaka, former section chief in Staff Headquarters of the Southern Army. Most of the information here regarding the *Giyūgun* and *Giyūtai* in Malaya therefore derives from interviews with Colonel Imaoka, from a written statement by Colonel Imaoka prepared for the author, and from an unpublished history by Colonel Imaoka, *Nansei hōmen rikugun sakusen shi* (Campaign history of the Army in the Southwest sector), 1965, in the Bōeichō Senshishitsu.

[5] Interview with General Inada, 1 December 1970, Tokyo.

Japanese officers assigned to train the *Giyūtai* were members of the *Dokuritsu Shubitai,* Independence Garrison Unit, under command of Major-General Shirataki Rishirō.The *Shubitai* strength throughout Malaya was four battalions initially; later it doubled in size. Members of the *Shubitai* travelled from village to town to city training the *Giyūtai* units. The *Dokuritsu Shubitai,* it should be noted, had no connection with independence policy, despite its name. The "Dokuritsu" in the name reflected rather its independence from divisional orders in the regular chain of command. It came directly under command of 25th Army Headquarters, later under the 7th Area Army. The *Dokuritsu Shubitai* had headquarters in Kuala Lumpur, Singapore, and Johore Bahru.

The main purpose of the *Giyūtai* was maintenance of peace and order, in effect a super-police unit. It was not intended as a battle formation. In March 1944 the total size of the *Giyūtai* was approximately 5,000. It was deployed in coastal defense positions and anti-aircraft operations in anticipation of Allied landings. The units engaged in construction of simple obstacles to coastal landing operations. Training of the *Giyūgai* continued until the end of the war. There was no plan initially for commissioning *Giyūtai* officers.

No Indians or Chinese were recruited into the *Giyūgun* or *Giyūtai,* Indians because they were instead channelled into the *F Kikan*-sponsored activities, and Chinese because they were not considered politically reliable. Both of these groups, nevertheless, were recruited into the *heiho* in Malaya.

One problem in the functioning of the *Giyūtai* was overlapping of duties with the Malay police. Another related problem was that young Malays preferred to enter the *Giyūtai* rather than the police because food and other material perquisites were more favourable in the *Giyūtai. Heiho* had even more volunteers because food and uniforms were supplied.[6]

One former *Giyūtai* member recalls having been confronted with the choice in 1943 of working on the notorious

[6] This was also Colonel Imaoka's statement. It should be noted, however, that most volunteers were given no choice, i.e., the alternative was the dreaded *rōmusha* work on the Thai-Burma Railway.

Thai-Burma Railway or joining the *Giyūtai.* He did not hesitate. He reports being trained in a camp near Muar in Malacca, where recruits were up at 6:00 a.m. for field drill. They made long marches, followed by running, followed in turn by jumping into a cold pool, as described by *Peta* trainees in Java. Often there was guard duty or drill again at night after a long day's march. They were given some basic language training, lessons in *seishin,* lectures on the Emperor, and practice singing *Kimi ga Yo,* the Japanese national anthem. The informant reported the Japanese commander of his *Giyūtai* unit was a courteous ex-diplomat rather than a professional soldier.

The same informant reports having attained the rank of first lieutenant and being sent to Singapore in the summer of 1945 with twenty-four others for military intelligence training. The original plan did not call for commissioned officers in the *Giyūtai.* He believes he was selected because his slight physique did not adapt him for work in the jungles. In Singapore they were taught espionage, the use of incendiaries, counter-intelligence, and sabotage. This training was under the charge of three Japanese officers who taught the Malay trainees for two hours every morning. For the rest of the day they were on their own. The training program continued for the last two or three months of the war. The trainee, unaware that the end of the war was imminent, recalls wondering why the Japanese instructors allowed them so much freedom.[7]

The Malay *Giyūgun* was created in early 1944 to assist in defense of all Malaya. Unlike the *Giyūtai,* it was organized and trained as a single unit with a central training camp and barracks at Johore Bahru. In April 1944, there were 2,000 men in the *Giyūgun,* a smaller and more disicplined unit than the decentralized *Giyūtai.* The Japanese Army provided uniforms and food in addition to military training and barracks. Weapons used in training were captured from the British Army. Japanese officers assigned to training the *Giyūgun* were from the same unit as for the *Giyūtai:* the *Dokuritsu Shubitai. Shubitai* officers involved in the training program were generally of first lieutenant or captain rank.

[7] Private interview, Kuala Lumpur, 17 February 1971.

Training was aimed at producing a fighting force, and the *Giyūgun* did go into combat together with the Japanese Army in the last defense of Singapore. Battle training of the *Giyūgun* continued until the end of the war, with its number never varying much from 2,000. The relatively small size and more intensive training by contrast with the *Giyūtai* reflected the difference in objective. Another reason for the smaller size of the Malay *Giyūgun* by contrast with *Peta* in Java was that so many young Malays were recruited for labor on the Thai-Burma Railway (100,000 by one estimate) that there were not many young men left. Still another reason for the small size was that the Chinese half of the populace could not be drawn on, nor could the Indian sector. Creation of the *Giyūgun* and *Giyūtai* in Malaya was motivated and also limited by the acute labor shortage caused by the large demand.

There were no special qualifications for *Giyūgun* recruits except that they be regarded as sympathetic to Japanese occupation. This in itself disqualified much of the population. Since all Malays received some Japanese language instruction in the school system, this was true also of *Giyūgun* recruits. As there was no Japanese plan for independence for Malaya, the Japanese were careful to avoid encouraging nationalism or aspirations for independence. As with all Japanese training programs, there was some emphasis on *seishin.* The *Giyūgun* was deployed along the coast defensively in Malaya and Singapore.

The call for volunteers for the *Giyūgun* went out in newspapers throughout Malaya in January 1944. The *Penang Shimbun* carried the announcement on 8 January, stating that the "Malay *Giyugun* is a regular force which will be stationed in Malay and will assist the Nippon Army in the various defense measures being carried out. They will not take part in expeditions to crush the enemy." Eligibility extended to all between the ages of sixteen and thirty-five who were physically fit and of "good character". It was announced that rank, salary, and promotion would be as in the Japanese Army, and that food, lodging and uniforms would be provided. There was the further exhortation that "youths imbued with enthusiasm for the cause of their freedom should speedily enlist without delay".[8]

The Malay *Giyūgun* included members of the banned *KMM* and some Indonesians. Ibrahim as Commander-in-Chief of the *Giyūgun* was given the rank of Lieutenant-Colonel. He referred to the *Giyūgun* as *PETA,* hoping to strengthen ties with its Indonesian counterpart. By early 1944 some in the *Giyūgun* made contact through Chinese with the MPAJA.[9] The Indonesians were in the same battalion, commanded by Indonesian officers. At the time of surrender, the Japanese made an agreement with an Indonesian commander, Brigadier-General Karim Rasjid, to send the Indonesian volunteers back to Indonesia.[10] Some *Giyūgun*-trained officers are in the present Malaysian Army. No doubt those who returned to Indonesia at the end of the war did so to elude the British on their return. In Indonesia, they joined the war of independence and were given ranks in the Indonesian Army. Some of them were given ambassadorial appointments and they generally had brighter prospects than if they had remained in Malaya, where many who "collaborated" are at pains to hide their wartime activities. Those Malays who served in the *Giyūgun* or *Giyūtai* have been notably less willing to reveal their wartime experiences than, for example, Javans or Burmans who served in Japanese-trained forces. The presumption is that public knowledge of this wartime activity would be a liability to Malays in public careers today.

In July 1945 the *KMM* group formed *KRIS,* acronym for a name often given as *Kesatuan Ra'ayat Indonesia Semenanjong,* or the Union of Peninsular Indonesians. Ibrahim and Dr. Burhanuddin jointly led *KRIS. Kris* is also the name of the short sword traditionally carried by Malays for self-defense. The plan was to raise the red and white flag of Indonesia on 8 August, the day Sukarno and Hatta arrived in Saigon to discuss Indonesian independence with General Terauchi. *En route* back to Indonesia, Sukarno and Hatta

[8] *Penang Shimbun,* 8 January 2604, *i.e.,* 1944.
[9] Soenarno, "Malay Nationalism", p.23.
[10] Gregory, Ann, "Dimensions of Factionalism in the Indonesian Military," mimeo. paper for the American Political Science Association, 1970.

because of essential petroleum resources.

There were a few *tokumu kikan* operating in Sumatra for specific goals whose staffs included *Nakano Gakkō* graduates. One example was the *Palembang Kikan.* Headed by a staff officer of the 25th Army who was simultaneously with the 9th Division of the Air Force, its staff included eight *Nakano Gakkō* graduates who trained one hundred Sumatrans. Sumatrans were divided into three groups and trained for periods of six months each. Ten *Kempeitai* also participated. All Sumatran trainees were armed with revolvers. Their primary duty was to protect the oil refineries of Palembang. Later, due to the impossibility of shipping oil to Japan, *Kikan* members instead performed some intelligence and counter-intelligence functions. At the end of the war most of the Sumatran members went to Java, where they joined the war of independence.[15]

Japanese policy toward Atjeh State was distinct from treatment of the rest of Sumatra. For one thing Atjeh State in northwest Sumatra had enjoyed a separate status for several hundred years. Restive under Dutch colonial control, the Atjehnese fought a war of independence from 1875 to 1908, attempting to regain their independent existence. IGHQ in Tokyo recognized the historic position and pride of the Atjehnese when General Sugiyama ordered Major Fujiwara as part of his intelligence mission in 1941 to establish contact with Atjehnese and to encourage pro-Japanese and anti-Dutch sentiment there.[16]

When Fujiwara met Chief of Staff Suzuki Sosaku of the 25th Army, Fujiwara was warned that he could not expect help from the 25th Army Staff, as they had no notion of a channel of approach to the Atjehnese, and it was furthermore not their duty. There were no such ready contacts for the Atjeh project as Fujiwara found in the Indian

[15] From an interview with N. Kurokawa, former *Palembang Kikan* chief, in Tokyo, August 1973.

[16] The information in this section on the activities of the *F Kikan* and Atjeh is derived from Nakamiya Gorō, "Sumatra ni okeru muketsu senryō no kage ni," (Behind the bloodless occupation of Sumatra), in *Shūkan Yomirui, Nihon no himitsu sen,* pp.93-7; and from Fujiwara Iwaichi, *F Kikan,* pp.273-5; 303-4; 312-14.

Independence League and Pritam Singh, at least not apparent to the 25th Army Staff.

At the end of 1941 the *F Kikan* project of enlisting Indian volunteers from behind British lines was going well. Lieutenant Nakamiya and a Japanese civilian assistant attached to the *F Kikan* went underground to Penang Island. When the two men were drinking in a bar one evening they were approached by two young men. They were Atjehnese from Kedah State. One of them, Sahid Abu Bakar, was a teacher in a Muslim school. Both told Nakamiya they wanted to fight to liberate Sumatra. They believed that if Indians and Malays could join with Japan to fight for independence, Sumatrans could do the same. "Our idea is to get rid of Dutch control to protect the Muslim faith and the nation's freedom," they explained.[17]

This was just the contact Nakamiya was looking for. He was also aware that the Atjehnese in Malaya were as anti-Dutch as the Atjehnese in Sumatra. There was, furthermore, he discovered, a Muslim group in Sumatra called *Pusa,* composed of Atjehnese. *Pusa,* acronym for *Persatuan Ulama Seluruh Atjeh,* or Central Organization of the Ulamas of Atjeh, was an anti-Dutch resistance group under religious leadership.

Just then a long-time Japanese resident of Malaya, Masabuchi, approached Fujiwara with another Atjeh contact. This man, Mohammad Saleh, had the same aim as Abu Bakar. The three Atjehnese patriots then joined the *F Kikan* and began propaganda broadcasts beamed to Sumatra. Abu Bakar, leader of the group, collected twenty Atjehnese from Malaya and assembled them at *F Kikan* headquarters in Ipoh for military training. Masabuchi worked with them along with the officers in charge of their training. The Atjehnese, watching the Indians in the Indian National Army, grew restive in Malaya. One day Abu Bakar approached Nakamiya and asked that the Atjehnese be allowed to infiltrate Sumatra, where they would contact *Pusa* headquarters and incite revolt in anticipation of the advance of the Japanese Army landings. A plan was accordingly concocted by the *F Kikan* and the young

[17] Nakamiya, "Muketsu senryō", p.94.

Atjehnese to infiltrate Sumatra to launch the struggle against the Dutch.

Main points of the plan were:

1) to land in north Sumatra and spread pro-Japanese feeling in the area;

2) during the Japanese Army invasion of Sumatra to protect transportation, communications, oil refineries, oil fields and bridges from destruction by the Dutch Army;

3) to procure food, water and information for the Japanese Army; and

4) to assemble abandoned Dutch weapons.

Fujiwara took the plan to Chief of Staff Suzuki of 25th Army Headquarters, and the plan was approved.

Selangor was the departure point selected for the Atjeh group, and the date was 7 January. But both Selangor and Kuala Lumpur were still in British hands. Nakamiya and Tashiro were charged with preparations for taking the group into Selangor. While they were making their way under cover to the coast, they hid one night in a building on a British rubber plantation. In the middle of the night, Nakamiya was awakened by a trembling Malay. Nakamiya discovered that their hiding place was surrounded by anti-Japanese Chinese workers. He decided to set fire to the building and, taking advantage of the ensuing confusion, the *F Kikan* group dashed to a river, where they submerged themselves and eluded the Chinese.

Another problem was the political stance of the government of the State of Selangor. Nakamiya met with the Malay state and village authorities to urge them to cooperate with Japan to free Malaya. Fujiwara's efforts with the *KMM* helped in the crisis, and the State of Selangor opted for cooperation with the Japanese Army.

On 16 January, six men in two small boats left Kuala Selangor with provisions for ten days, plus bombs and grenades. Nakamiya watched them disappear heading across the Straits of Malacca. A second group of six left southern Selangor on 25 January headed for the east coast near Medan. It was two months before the *F Kikan* learned what had become of the Atjehnese patriots who had headed across the Straits in their small boats.

A call from 25th Army Headquarters in early March informed Nakamiya that a secret mission had reached Penang from Sumatra. Nakamiya was delighted to learn that eight men in two groups had escaped from Sumatra on 3 February, arriving in Penang on the 20th. According to the spokesman of the mission, Abdul Hamid, Abu Bakar's group had landed safely at Sungei Sunbifun and Bagan Siapiani. But, spotting Dutch coastal guards, they abandoned all their weapons and landed in the guise of refugees. They were questioned for two days by the Dutch in the Medan jail, during which time they managed to smuggle a letter out to the *Pusa* leader. For lack of proof and because of tactics used by *Pusa,* Abu Bakar and his group were released without having revealed any information. They immediately set to work collecting men.

As soon as *Pusa* members got wind of the Japanese Army's advance into Malaya on 8 December, they began speeches and propaganda to foment national rebellion. In mid-December *Pusa* leaders met to vow loyalty to the Muslim nation and cooperation with the Japanese in the struggle against the Dutch. They planned a series of *Pusa*-instigated riots. Japanese propaganda broadcasts from Malaya and the arrival of the Abu Bakar group heightened their spirit of resistance. The more oppressive Dutch tactics became, the greater the tension among the Atjehnese.

On the night of 23 February a riot broke out, during which the Dutch Civil Administration Office at Seulimeum was attacked. By the next morning, three hundred armed men wearing the "F" insignia of the *F Kikan* removed Dutch dynamite fuses on twenty bridges. Five men were killed when a bridge exploded before they could defuse the dynamite. Other "F" members, using sabotage against the Dutch Army, cut telephone lines and destroyed railroads. *Pusa* leaders waited impatiently for the arrival of the Japanese Army to tip the scales in the Dutch-Atjehnese struggle.

On 11 March, the *Konoye Shidan,* Imperial Guards Division, left Singapore for Sumatra, crossing the Malacca Straits. As they approached Kotaradja, Masabuchi and Nakamiya, accompanying the Japanese landing force, spotted men in small boats wearing red "F" insignia on their arms. They had come to welcome the Japanese invasion force. Along the coast, Atjehnese appeared with rice bowls, fruit and wine. Cars and bicycles were also ready and waiting. Nakamiya was deeply impressed. March 12 marked a bloodless landing, thanks to the Atjehnese and the work of the *F Kikan.*

Nakamiya learned that six days before the landing a proclamation had gone out to all parts of Atjeh State that Atjeh would revolt at once against the Dutch and cooperate with the Japanese. *Pusa* demanded that the Dutch Resident return the government of Atjeh to the Atjehnese. On the night of 11 March, the Atjehnese revolted and began fighting the Dutch in Kotaradja. The Dutch Army attempted to set fire to airports and oil storage depots, and "F" members fought to prevent the destruction. At 4:00 a.m., there was a general attack on Dutch officers and barracks in Kotaradja. The arrival of the Japanese added to the confusion among the Dutch, whose resistance collapsed in face of the joint Atjehnese-Japanese attack.

Part of the enthusiasm of the Atjeh welcome to the Japanese Army was abetted by the chance coincidence that Sumatrans mistook the red and white insignia of the Japanese expeditionary force for the flag of independent Indonesia. This coincidence encouraged hundreds of Atjehnese to volunteer for the *F Kikan.*

The *F Kikan* mission succeeded with the Atjehnese as with the Indian independence movement in Southeast Asia. With the dissolution of the *F Kikan* in March and its reorganization as the *Iwakuro Kikan,* the Atjeh project was officially completed. Masabuchi, however, remained to work in Atjeh state until the end of the war, when he committed suicide.[18]

Apart from the Atjeh project, the Japanese Army in Sumatra as in Java and Malaya created a *Giyūgun* and *heiho* units. Part of the impetus for organizing the *Giyūgun* was the same as in Java: to shore up deficient troop strength and supplement the Japanese Army. As in Java,

[18] Fujiwara, *F Kikan,* pp.313-14, and interview, July 1972, Tokyo.

the *Giyūgun* was created after the September 1943 order from Southern Army Headquarters. Though in Java it was Japanese policy not to recruit ex-*KNIL* officers or men, in Sumatra, the 25th Army adopted a different attitude. On the assumption that the cultural and educational level in Sumatra was lower than in Java, the decision was made to recruit men from Menado and Amboina who had been in the *KNIL*, who were educated in Dutch and English, and who could therefore be used as interpreters. Ambonese and Menadonese, in fact, formed the nucleus of the officer corps of *shotaichō* in much of Sumatra. In Atjeh, however, because of the bitter hatred of the Dutch and all who cooperated with them, it was impossible to consider using people from Amboina or Menado, who were not only Dutch-educated but Christian as well. Instead, many *Pusa* members were recruited into the *Giyūgun*.

Several officer training centers were established for the *Giyūgun* at Medan, Padang, Kotaradja, and Palembang. In March 1944 there were some thirty companies (*chūtai*), and at the end of the war the size of the *Giyūgun* reached between five and six thousand, or roughly the size of the *Giyūtai* in Malaya.[19] An estimated half of these units were trained in Atjeh.

The Sumatra *Giyūgun* was organized and trained at the state level, not in a single unit under unified command as at Bogor. Another difference is that there were almost no *Nakano Gakkō* graduates involved in training the Sumatra *Giyūgun;* nor was there a large *tokumu kikan* operation involved. Officer ranks were also lower than in *Peta.*

Initially there were at least three centers for officer training, possibly as many as six. The training was based on methods used in the Japanese Army but within an abbreviated three months' period. Candidates selected were sons of Muslim leaders, officials and teachers, with the exception of the Menadonese and Ambonese. The selection process was designed to select out the educated strata. The officer candidates underwent a rigorous daily regimen beginning at 6:00 a.m., as was true of all training programs

[19] Again, most of the information here, unless otherwise indicated, has been kindly supplied by Colonel Imaoka, orally or in writing.

in Southeast Asia. The curriculum included military drill, judo, *kendō,* Japanese language, and *seishin* lectures. The propaganda emphasis was on fighting for one's country and cooperation with Japan within the Greater East Asia Co-Prosperity Sphere. It was impossible for Japanese officers to discuss directly independence, which was the known objective of at least the Atjehnese. After three months, officer candidates became *shotaichō,* or platoon commanders, which in the context of Japanese control of Sumatra was a non-commissioned rank equivalent of sergeant. Simultaneous with the training of Sumatran officer candidates, Japanese officers and non-coms were also given training in how to handle Indonesians and in the Indonesian language. Their training lasted approximately two months.[20]

The principal duty of the *Giyūgun,* by contrast with the *heiho,* was coastal defense. The training was therefore designed to produce officers and men ready for some combat duties. Though the *Giyūgun* was never called on to fight (until the war of independence) it was prepared to do so. Rifles captured from or left by the Dutch were used during the training. While in Malaya the duties of public peace operations and coastal defense were generally divided between *Giyūtai* and *Giyūgun,* in Sumatra they were combined. Japanese units involved in training the *Giyūgun* were at the early stage the *Dokuritsu Shubitsi* and later the *Konsei Ryōdan,* or Mixed Brigade. The *Dokuritsu Shubitsi* in Sumatra was headquartered at Sibolga in the north and Lahat in the south. The 3rd and 4th Regiments of the *Konoye Shidan* at Medan also took part in training programs.

The Sumatra *Giyūgun* lacked a distinctive name such as *Peta* and did not have the sense of identity or esprit of *Peta. Giyūgun* units in Sumatra were attached directly to units of the Japanese 25th Army rather than being organized in battalions under Sumatran commanders.

When officer training was completed, the officers were

[20] Satō Morio, an Indonesian language specialist who participated in the training, states that these training programs took place in Medan the late 1942. This seems too early according to other accounts: interview, July 1972, Tokyo.

dispersed with Japanese advisers in several small units throughout each state to recruit and train men. For Atjeh, for example, the officer candidates from Kotaradja and other parts of the state were assembled at the Medan training center with candidates from other states. Following the three months' training period, they dispersed to sixteen places throughout Atjeh to train recruits. Each of the sixteen basic officer units in Atjeh consisted of six Japanese and five Sumatrans: one Japanese officer, two non-coms, and three soldiers; and two Sumatran non-coms and three soldiers. At these sixteen points in Atjeh volunteers were recruited locally from among the most educated classes. While some may have been college graduates, others had not graduated from primary school, depending on the local educational level. Sons of Muslim leaders and *Pusa* members were notable among the recruits. Local training of *Giyūgun* volunteers continued until the end of the war. During the harvest season they were given time off to go home to help harvest the rice; in this way they contributed labor and helped raise rice production. The same basic training system was used in each state throughout Sumatra.[21]

The Japanese in Sumatra made some conscious use of "divide-and-rule" tactics. In late 1942 a plan was drafted for organization and training of a *Keibitai* (police unit) of young Atjehnese to be used both in Sumatra and Malaya to preserve peace and order among the Chinese. The Japanese recognized the martial spirit and tradition among the Atjehnese and hoped no doubt to vent it against people other than themselves. The people of Atjeh and the Minangkabau were regarded by the Japanese as of higher capabilities than other Sumatrans.[22]

Satō Morio and others involved in training the *Giyūgun* were well aware of the significance of Islam in the life of

[21] Satō Morio believes the total size of the *Giyūgun* was not over 2,000. This estimate of Satō's is much lower than estimates by Kanahele and Imaoka. Kawadji Susumu and other *chūtaichō* involved in the training believe the total size was between five and six thousand: interview 4 August 1973, Tokyo.

[22] *Tokugawa shiryō* (Tokugawa materials), no.28, Reference materials on nationality policies, November 1942, marked "Sōmubu sōmuka" (General Affairs Bureau, General Affairs Section).

the Sumatrans. He met regularly with Muslim leaders to explain the problems and requirements of the Japanese Army in an attempt to elicit their support of the training programs and other Japanese requirements. It is interesting to note that, despite Japanese support of Islam as a matter of policy, there was no attempt made by the Japanese to exploit the Muslim concept of *Jihad,* or Holy War on Sumatra, though on Java Professor Benda considers the identification of Japanese and Islamic holy war the most important success of Japan's Islamic policy.[23]

The Japanese attitude toward Sumatra during the war was tinged with romanticism. Apart from its wealth of natural resources and natural beauty, Sumatra appealed to the Japanese as a kind of sparsely populated tropical utopia. Rumors that verged on the fantastic were current toward the end of the war. One was a report that in 1945, IGHQ actually considered moving to Sumatra when Tokyo was under heavy fire. Another was the story of a contingency plan of General Muto to bring the Imperial family to Sumatra to ensure their safety. Muto allegedly approached Satō Morio and asked him to reconnoitre and report on an appropriate site.[24] And some individuals in the Japanese Army contemplated personally retiring to Sumatra after the war.

There were also some small Japanese units created at the end of the war. Late in 1944, a Japanese guerrilla unit or *yūgekitai* was formed. In Atjeh, it consisted of some thirty Japanese officers, including some graduates of the *Nakano Gakkō*. Satō recalls teaching them Indonesian. Their main function was intelligence, despite their name, and though they could not hide their Japanese identity, they were disguised as Sumatrans. They went about with long hair and no uniforms, selling tobacco, helping fishermen, and harvesting rice as they collected intelligence.

One small guerrilla unit of Sumatrans was the *Zanchi Yūgekitai,* or Remaining Guerrilla Unit. It was trained beginning in May 1944 by the *Konoye Shidan* on orders

[23] Benda, Harry J., *The Crescent and the Rising Sun,* p.141.
[24] Interview with Sato, 5 June 1972, Tokyo.

from the Seventh Area Army, already anticipating Allied counter-invasion. The plan envisaged a guerrilla unit of approximately four hundred men who would be trained by twenty Japanese officers. The unit was expected to operate in the agricultural villages to prevent a return of the Dutch. The training manual was prepared by Kondo Tsugio, a graduate of the third class of the *Nakano Gakkō,* sent to Sumatra from Singapore in January 1943.

Kondo organized a small *tokumu kikan* called the *Kondo tai* in Medan in 1943. His was essentially an intelligence and reconnaissance mission to support military operations. There were not even any adequate maps of Sumatra available to Kondo at the time. In his office were five Sumatrans whom he trained in intelligence techniques. Kondo reports he trained these men to organize and arm groups in the *kampongs,* and that his unit was the nucleus of a force of 50,000 at the time of the war of independence.[25]

Four graduates of the *Nakano Gakkō* were sent to Sumatra in early 1942 on individual intelligence missions. They were attached to the state police in Atjeh, the West Coast State, the East Coast State, and Army Headquarters. Among these men was Adachi Takeshi, one of the civilian language specialist graduates of the third class of the *Nakano Gakkō.* Adachi had studied Chinese in the *Nakano Gakkō,* but as there were not enough English-trained *Nakano Gakkō* graduates for all the intelligence missions in Southeast Asia, the Army resorted to sending some Chinese-trained graduates to Southeast Asia instead. All these men operated in small groups or individually rather than in a single large *tokumu kikan.*[26]

Satō Morio also alleges to have organized a small *Satō Kikan* in Atjeh, following completion of his assignment in the *Giyūgun* officer training program. His *Kikan* was created on order of the *Konoye Shidan* commander, General Muto, who was concerned about anti-*Kempeitai*

[25] Interview with Kondo Tsugio, 11 April 1972, Tokyo. Kondo may have been speaking of the combined impact of the *Giyūgun* and *heiho* troops with guerrilla units as they merged in the war of independence.

[26] Interview with Adachi Takeshi, 11 April 1972, Tokyo.

sentiment and activities. Its chief function was intelligence. The *Satō Kikan* was divided into several small units, each headed by a Japanese officer, under whom were two or three non-coms and a few soldiers. Information was bought from Sumatrans, according to Satō. The *Satō Kikan* was organized after completion of the *F Kikan's* project in Atjeh and had no connection with it. Satō reports the most difficult aspect of his whole assignment in Sumatra, whether with the *Giyūgun* or *Satō Kikan,* was to elicit the understanding and support of the Sumatrans, who were so different in character, culture and tradition from the Japanese. Every good Muslim prayed five times a day, and soldiers in the midst of their training schedule showed no hesitation in interrupting their training for prayer. This infuriated their Japanese instructors. In Atjeh more than elsewhere in Sumatra there were numerous anti-Japanese incidents and clashes between Sumatrans and Japanese troops. The Atjehnese passion for independence led them to shift the target of their antagonism rather rapidly from the Dutch to the Japanese after occupation. Yet there was no *Giyūgun* or general revolt of the magnitude of those in Burma or Java. The training programs were not interrupted until the end of the war.

At the end of the war in Sumatra, violence erupted against the Japanese, the returning Dutch, and between groups of Sumatrans. In Atjeh State, for example, members of *Pusa* killed many of the *ulèëbalang,* the aristocratic class through whom the Dutch had ruled the State of Atjeh. Japanese authorities were unable to control the situation, and though there were numerous clashes between Sumatrans and Japanese, there were also Japanese soldiers who fought in the war of independence on the Sumatran side. Some of these men have remained in Sumatra after the war.

In North Borneo, the Japanese Garrison Army, under command of General Yamawaki Masataka, created in 1944 a small *Giyūgun* of approximately 1,300 men. One battalion of about 500 men was stationed at Kuching. Four other areas – Miri, Jesselton, Sandakan, and Sebau – each had one company of from 150 to 200 men. Their duties were preservation of public peace and order plus some defense and guard duty. They also collected intelligence. Recruits were of the Dayak ethnic group.[27]

Until some time in the later summer of 1943, Japan felt secure in French Indochina because of the 1941 defense treaty with the Vichy French régime. In Indochina, Vichy officials under Admiral Decoux kept De Gaullists in prison and maintained the terms of the treaty agreement with Japan under which Japan stationed some 35,000 troops in Indochina.

In late 1943, the situation began to change. When U.S. forces landed on Leyte it appeared that American invasion of Indochina was imminent and might threaten Southern Army Headquarters. Burma, Malaya, and Thailand were at the same time under British threat from the West. Because of this combined military menace, De Gaullist sympathies increased among Frenchmen in Indochina. De Gaullism burgeoned, in fact, to the point where anti-Japanese incidents instigated by French residents or by De Gaullist sympathizers erupted among the Vietnamese. French authority seemed to sanction this trend when French police also cracked down on pro-Japanese nationalists in 1944. Destruction of the Japanese flag, overturning Japanese transport trucks, and arrests of Annamese friendly to the Japanese occurred with increasing frequency.

These changes prompted the Japanese to plan a settlement in Indochina to redress the situation, restoring control to Japan and at the same time supporting a Vietnamese régime and the French administrative structure. French authority was to be disestablished while preserving the French administrative structure. The plan was known by the code name "Operation Ma", later "Operation Mei".

Discussions took place in IGHQ during January 1944 on what form the settlement should take. An important variable which the Japanese had to consider was the existence of the Vichy Régime. So long as the régime remained secure it was deemed advisable to utilize the existing French administrative structure; but this could no

[27] Imaoka, *op. cit.*, pp.160-1.

longer be judged constant.[28] The decision to disestablish French authority in order to preserve Japan's military structure therefore seemed justifiable.

After the Allied landing in Normandy on 6 June 1944, Japanese Ambassador Yoshizawa cabled the Greater East Asia Ministry for instructions on Japanese policy when the Vichy Government suspended activities. On 14-15 September, the Supreme War Guidance Council devised contingency plans in anticipation of a sudden change in the situation. Though Foreign Minister Shigemitsu suggested independence for Indochina, he was voted down because of the possibility of military resistance.[29] But the decision for a military settlement was postponed until February 1945 because of Japanese military engagement in the Philippines. Southern Army Headquarters was moved from Manila to Saigon in November 1944. The Japanese Army in Indochina was reorganized as the 38th Army in December.

Southern Army Staff Headquarters had devised plans for control of Indochina after the "Mei-go" Operation. On 31 October, Chief of Staff Kawamura of the Indochina Garrison (which became the 38th Army in December 1944) ordered a *Kempeitai* Lieutenant-Colonel to create a plan for control and submit it by mid-December. The *Kempeitai* officer felt there was no need to impose a military government during a military emergency, but rather that maximum local support could be elicited by liberation from French control and by recognizing "independence". Within 38th Army Headquarters many objections were raised to the suggestion of independence rather than military government. Parts of the plan were nevertheless utilized and approved by Kawamura and the Supreme War Guidance Council.

In actuality, Japan did not envisage either genuine independence or a military administration. Japanese control was imposed over the French administrative structure. But this Japanese stance encountered two problems: reluctance of French officials to serve under Japanese command, and

[28] Bōeichō Bōeikenshūjo Senshishitsu, *Shittan Mei-go sakusen* (The Sittang and "Operation Mei"), p.582.

[29] *Ibid.*, p.584.

realization by some Vietnamese leaders that Japan did not contemplate genuine independence.

Japan determined to support Emperor Bao Dai and a cabinet of his choice. This step was designed to solve these problems and to create the illusion of independence and the substance of real cooperation with Japan simultaneously.

As part of the plan to elicit Vietnamese support after the "Mei-go" operation, in January 1945 a small *tokumu kikan* was created, the inspiration of Southern Army Staff intelligence officer Lieutenant-Colonel Tada Minoru. Called the *Yasu Butai* or Annam Unit, it was placed under command of Southern Army Headquarters and headed by Lieutenant-Colonel Ishida Shoichi. Although the *Yasu Butai* was essentially an intelligence agency, it was exceptional as a *tokumu kikan* in that Ishida was not a *Nakano Gakkō* graduate. Under him, however, were a major and two captains who were *Nakano Gakkō* products, and twenty second lieutenants, plus a few Japanese and Vietnamese civilians. Headquartered in Saigon, the *Yasu Butai* had branches in Hanoi, Hué, Haiphong, and Pnom Penh.

The duties of the *Yasu Butai* were to destroy French forces in Indochina, to create disturbances within the French Army, to capture French military facilities and weapons, to carry out counter-intelligence, to foster pro-Japanese sympathy within the Vietnamese government, and to encourage the Vietnamese independence movement. After its creation the *Yasu Butai* was transferred from Southern Army Headquarters and placed under command of the 38th Army. Both Headquarters, it may be noted, were in Saigon after December 1944. The *Yasu Butai* was organized and operating two months prior to the start of Operation "Mei-go".

Creation of the *Yasu Butai* was a much later phenomenon than the organization of other *tokumu kikan* in Southeast Asia. The reason for the timing was that Japan was not anxious to see Vietnam become a battleground, so long as the Japanese-Vichy treaty was observed and the *status quo* not threatened.

Planning for the formation of the *Yasu Butai* began in a back room of 38th Army Staff Headquarters on the night of 10 January 1945. Organizers of the *Yasu Butai* were Colonel Ishida, Captain Kaneko Seigo, Major Sasahara, Major Fukube, Captain Ishikawa, and Hayashi Shucho. The unit was to be part of a total campaign in Vietnam, the "Mei-go" operation. Insignia of the unit was the Japanese flag with the character "*An*" or "*Yasu*" in the center, symbolizing the joint struggle of Japan and Annam.[30]

A three-phase operation was conceived in the strategic planning sessions of the staff officers. The first phase included gathering intelligence on the French Army, collecting local intelligence reports, propaganda, support of the Annam monarch, strengthening underground organizations dedicated to Vietnamese independence, and contact with French officers of the Pétain faction.

A second phase of strategy was to be coordinated with the military "Mei-go" offensive. This phase included surveillance of Bao Dai, supporting the political power of the Vietnam independence party, preservation of peace and order, strengthening the intelligence network, forming a volunteer coastal unit in anticipation of Allied landings, and preparations for guerrilla warfare in the mountains and forests.

Independence movements, as elsewhere in Southeast Asia, were to be strengthened insofar as commensurate with fostering sympathy with Japan. Because of Japan's experience in India, Malaya, Java, and Burma, this was regarded as the Achilles' heel of the whole operation. It was psychological warfare, and *Nakano Gakkō* men were the specialists assigned to the project.

To carry out the project Kaneko, with Lieutenants Hidaka and Tamimoto on 6 February crept into the coastal town of Hué in central Indochina. They rented a store and operated as merchants. There they reconnoitered the geography, gathered intelligence on French installations and met secretly with the members of the Vietnamese Independence Party. They captured valuable documents, including a roster of all espionage agents in Indochina. They

[30] Information on the formation and activities of the *Yasu Butai* comes from an article by Kaneko Seigo, "Annam himitsu butai", (Annam secret unit) in *Shūkan Yomiuri, Nihon no himitsu sen,* pp.161-3.

used wire-tapping to monitor important conversations of French officials. Young boys from the Annam Independence Party were persuaded to cooperate in the interests of freedom for their homeland. Those who volunteered were transformed into salesmen, boatmen, priests and photographers. One beautiful young woman collected intelligence from a river sampan. A young Vietnamese eluded a pursuing intelligence officer by hiding for several hours in the murky waters of the palace moat at Hué.

The date for a military settlement was fixed for 9 March 1945. The plan for a military coup was contingent on Admiral Decoux's rejection of Japanese demands. It was decided to support independence aspirations, at least in form. In March the new Japanese Ambassador Matsumoto Shin'ichi accordingly presented Admiral Decoux an ultimatum demanding that he agree to placing the French Army, Navy, police, administration and banks under Japanese command. Decoux refused as anticipated, and the coup took place 9 March. Bao Dai was accordingly confirmed in his authority as Emperor.

On 9 March, Japanese forces moved to take over all government offices and garrison camps. The whole French community was disarmed and interned. One gambit of the *Yasu Butai* was the personal surveillance of Emperor Bao Dai and reconfirming him in his sovereign rights over Annam and Tonkin. On the night of 9 March members of the *Yasu Butai,* dressed as Annamese, carried out the plot. Inside the palace were hiding other *Yasu Butai* members who had entered it during the day disguised as sightseers. They hid when the gates were closed in the evening. It was their duty to reopen the gates on a signal. Simultaneous attack was made against the French Army barracks. Kaneko and his interpreter were to enter the palace grounds by sampan and spirit the Emperor and Empress away. But the boatkeepers, confused by the sound of gunfire and the blare of a *Yasu Butai* megaphone, refused to cooperate, and Kaneko had to carry out his mission on foot. He finally encountered the Emperor and Empress, elegantly dressed and in an ornate carriage. Through his interpreter Kaneko said to the royal couple:

"The action of the Japanese Army is part of the fight for Annam's freedom and independence. The long period of French colonization is now completely ended, and after tomorrow it will be Annam for the Annamese, an achievement of brilliant glory!" Kaneko reports the Emperor wept, and, shaking hands with Kaneko replied, "I'll never forget your words. After the war we shall surmount our sufferings and cooperate with our friend, Japan. Today is a memorable day for me."[31]

On 10 March the Japanese Ambassador in Saigon declared Vietnam independent. Between that day and the end of the war, the *Yasu Butai,* headquartered now on the Annam-Laos border, carried on guerrilla warfare in the forests and mountains of Annam against the French and against Britain's Force 136, which operated also in Thailand and Malaya as an élite commando and intelligence unit. The coastal defense volunteer unit was deployed along the coast in anticipation of Allied landings.

The plan for control of Indochina after the "Mei-go" offensive called for creation of a small local armed force, a *Giyūtai. Giyūtai* in Southeast Asia were generally more limited in conception than *Giyūgun.* The assumption was nevertheless that an independent nation needs an armed force. As soon as Vietnam, Laos and Cambodia became independent they would need an army. Japan should therefore help to recruit, organize and train a volunteer force, though there were no weapons available for a training program. The *Giyūtai* would help to maintain public peace and order for the duration of the war and for self-defense afterwards.

The plan called for organizing five units of one thousand men each in Tonkin, Annam, Cochin China, Laos and Cambodia. Recruitment was to be through physical and written exams based on those used in Japan, at least in the ideal. Training was to be under auspices of operations staff officers of the 38th Army, and the training therefore took place after the December reorganization of the 38th Army. The *Yasu Butai* was not involved in the training of the

[31] As related by Kaneko, "Annam himitsu butai", p.163.

Giyūtai. There was no provision in the plan regarding ranks of indigenous officers or for having the *Giyūtai* engage in actual combat together with Japanese forces.[32]

In May 1945 an announcement was finally issued calling for volunteers. Organization of the units in Tonkin, Annam, and Cochin China began in June and training was under way in July. In Laos and Cambodia, however, recruitment exams were still not completed in August when hostilities ceased. The timing of the organization of this *Giyūtai* makes it therefore impossible to judge the effectiveness of the unit. It also means that in 1945 there were no Vietnamese-controlled combat forces other than those under Ho Chih-minh.

The Japanese Army also trained a special unit called the *Tokushu Kōyōnin* (special employees) who were used as sentries at military installations. There were one hundred of these guards in Saigon. Difficulties arose between the sentries and the police, who were under the jurisdiction of the French Governor-General. The problem was brought to diplomatic negotiation at one stage. This small unit was used until the end of the war.

Training of units in Indochina, because it was so belated, was not as effective as in Sumatra and Malaya. It is not really possible, because of the brevity of the training, to ascertain any real impact on post-war Vietnam.

The Philippines under military occupation presented the Japanese with a critical problem in popular resistance, in part through underground guerrilla activities. The Japanese attempted in several ways to solve the problem and to mobilize political support for the military administration policies. They never really succeeded.

The Japanese at the end of 1942 abolished all existing political parties and in their place created a supra-party, the *Kalibapi* (short for *Kapisanan Sa Paglilingkod Sa Bagong Philipinas,* Association for Service in the New Philippines). It was headed by chief executive Vargas as president *ex-officio* and Benigno Aquino as vice president, both

[32] Much of this information is derived from an interview with a former *Kempeitai* Lieutenant-Colonel, June 1972, Tokyo; and a statement by and interview with Colonel Imaoka, 15 March 1972.

professing support for the goals of the Co-Prosperity Sphere. *Kalibapi* was organized at the local level and had attached to it a Junior *Kalibapi* and Women's Auxiliary. The government boasted in 1944 that it had 1,500,000 members.[33]

In May 1943, in their continued search for loyal support the Japanese organized *Saipilnip,* based on former residence in Japan. General Artemio Ricarte became president, with José Laurel II, the president's son, as secretary-general. The Japanese also attempted to attract discontented groups such as the agrarian radical Sakdalistas.[34]

Still another attempt was made to maintain control through formation of the Philippine Constabulary, charged with maintaining peace and order and in particular with coping with the guerrilla problem. In November 1943, the Japanese pressed Laurel to increase the size of the Constabulary to 40,000 and to establish regional training schools to improve the quality of members. Former officers of the Philippine Army were given automatic commissions to expedite the pacification program. Despite numerous awards and incentives issued by Laurel, however, the Constabulary never was effective from the Japanese viewpoint. The basic problem was that the Japanese Army was hesitant to arm the Constabulary, fearing the arms would reach the hands of the guerrillas. Guerrillas had the sympathy of most of the populace, who were hostile to Japanese military administration. The Constabulary was an old pre-war unit organized in 1901 for police purposes and led in the thirties by Maj.-Gen. Guillermo Francisco who was also Chief of Staff and Director of the Constabulary in wartime.[35] Before the war the Constabulary was underpaid, which encouraged its members to try to supplement their salaries through graft. Wartime allegiances shifted mercurially between the guerrillas and the Constabulary, which enabled the guerrillas to use double agents successfully. Physically, Constabulary members were easy to

[33] Steinberg, *Philippine Collaboration in World War II*, p.96.
[34] *Ibid.*, pp.96, 107.
[35] *Ibid.*, pp.58-9, 95-6. Francisco was replaced as chief of the Constabulary in August 1944 by Paulino Santos.

identify because of their clean uniforms, which made them prime targets of assassination by guerrillas.[36] Constabulary members, unlike the *Makapili*, were not tried for treason after the war, apart from a few individuals.

Japan did not intend, after granting the Philippines "independence" in October 1943, to create an Army and Navy there but rather to strengthen the police power in order to maintain public peace. The 14 November 1944 General Guidelines for the Luzon Island Campaign contained the statement, "We will make use of part of the armed strength of native units who have friendly feelings toward us. We will lead them in active guerrilla warfare at the same time that we carry out the campaign."[37] When MacArthur landed on Leyte on 19 November, the situation became critical for Japan.

Accordingly, on 8 December the Japanese combined all their efforts at support, creating the *Makapili* (abbreviation for *Kalipunang Makabayan, ng mga Pilipino,* the Patriotic League of Filipinos), under the command of General Artemio Ricarte. Estimated size of the *Makapili* was between 3,000 and 5,000.[38] Whereas the Japanese mistrusted the Constabulary, they deemed the *Makapili* loyally pro-Japanese and armed them as they could spare arms. The *Makapili* consisted in part of groups of Sakdal/Ganaps, an impoverished radical tenant farmer party which was encouraged through the efforts of its leader Benigno Ramos to look to Japan for liberation from bondage. There were also other local pressures to join, including the need for a counter-weight against the radical Hukbalahaps. The term *Makapili* had been used during Spanish rule to designate the guards of the Governor-General who were descendants of

[36] Phone interview with Wendell Fertig, wartime commander of guerrilla operations on Mindanao, 28 November 1973, Colorado.

[37] Written statement by Colonel Imaoka for the author.

[38] Colonel Imaoka's estimate of the size was 15-20,000. This represented the Japanese goal, not actual size. Steinberg's estimate is 5,000, *op. cit.*; and see Steinberg, "The Philippine 'Collaborators'; Survival of an Oligarchy", in Silverstein, ed., *Southeast Asia in World War II: Four Essays,* pp.67-78. Hartendorp also suggests the figure 5,000: *History of Industry and Trade of the Philippines,* p.138.

Yaqui Indians brought from Mexico by the Spaniards.[39] The Japanese hoped to recruit 20,000 men into the *Makapili* but did not succeed. Most recruits were from the Central Luzon provinces of Nueva Ecija and Bulacan and from the Southern Tagalog provinces of Rizal and Laguna. The training took place in these areas, and it was here that the *Makapili* fought most effectively.[40] These were also the same areas where the Huks gained major strength, and the *Makapili* must therefore be seen in the context of the complex struggle between guerrilla groups. General Ricarte, though nominally supreme advisor, was never enthusiastic about the *Makapili.*

Laurel, Japanese Ambassador Murata Shōzō, and 14th Army Chief of Staff General Muto were all opposed to the creation of the *Makapili,* but 14th Army Commander-in-Chief General Yamashita, overriding their opposition, approved it. Benigno Ramos became president, Pio Duran vice-president, and General Ricarte supreme advisor.

By the end of 1944 the 14th Army could spare neither officer personnel to train a genuine *giyūgun* nor enough weapons to arm it. Consequently the *Makapili* functioned more as *heiho,* which had not been organized in the Philippines. *Makapili* members helped dig trenches, served as guides in the jungles, and collected intelligence. Only east of Manila did they engage in combat, together with the 8th Division under command of General Yokoyama Shizuo.[41]

Ricarte turned instead to organizing a third unit, a *Giyūgun* or *Hoān Giyūgun* (security force) in October, 1944. This unit, though theoretically supported by funds from the Japanese through Ramos (who pocketed the money) was in fact more independent of Japanese control. Its existence was rather predicated on loyalty to Ricarte. Laurel looked more kindly on this *Giyūgun* than on

[39] Interview with Wendell Fertig, and correspondence with Professor Ron Edgerton. Professor Edgerton has provided several useful suggestions relating to the Philippines.

[40] Hartendorp, *Industry and Trade in the Philippines,* v.1, p.138.

[41] Information supplied by Maj.-Gen. Utsunomiya, former Deputy Chief of Staff of the 14th Army, to Dr. Yoji Akashi, November 1973.

the *Makapili* because of the relative independence from Japanese control. A Japanese plan for a coup and merger of the *Makapili* and *Giyūgun,* the combined forces to be put under the more pliant Ramos, did not come off because of opposition from Ricarte.[42] Near the end of hostilities Ricarte led his two or three hundred soldiers into the mountains, following Yamashita's group as they withdrew from Manila.

The *Makapili* were more clearly identifiable at the end of the war than the wartime Constabulary or *Hōan Giyūgun.* The *Makapili* were regarded by the majority of their compatriots as traitors; this judgement was formalized in a Philippine court ruling after the war, when they were tried for treason.

Japanese efforts at arming Filipinos also included organization of a *Shin'eitai* (Honor Guard) attached to Laurel's office. Its size was estimated at about 250 men, organized in two companies armed with machine guns and pistols. It was commanded by José Laurel III.

Japanese attempts to use Filipinos to control Filipinos basically failed because of local pressures and pre-war American colonial policy. Collaboration with Japan connoted not loyalty to Japan but the desire to gain or retain political power. American colonialism had encouraged rather than competed with Philippine nationalism. Independence had been promised as early as the 1934 Tydings-McDuffie Act. There was no contradiction between fighting for the return of MacArthur and for independence. Japan was not seen as a necessary link in the evolution toward independence. Japan's appeal as liberator was thus less effective than elsewhere in Southeast Asia. This made it difficult for the Japanese to mobilize troops against guerrilla resistance.

The Japanese occupation of the Philippines, however, as in the rest of Southeast Asia, gave an opportunity for men of military ability to rise. Unlike Burma and Indonesia, where there was a linear relationship between wartime Japanese training and participation in post-war military/

[42] Interview by Dr. Yoji Akashi with Mr. Ota, Japanese adviser of Ricarte from 1921 to 1945.

political élites, in the Philippines Japanese occupation had an inverse effect. It was those who fought and led the anti-Japanese guerrilla units, rather than the *Makapili* or Constabulary commanders, who rose to post-war political leadership. Foremost in this category were Ramon Magsaysay, Ferdinand Marcos, Ruperto Kangleon, Salipada K. Pandatun, and Macario Peralta. Magsaysay and Marcos have made use of their wartime guerrilla activities in recent political campaigns with success.

6

Revolt of the Independence Armies

As Japan's military weakness was increasingly exposed in Southeast Asia, a corresponding shift in attitudes of nationalist leaders toward Japan was discernible. The luster of Japan's early military successes was dimmed by her inability to defend the areas she had occupied initially, as well as by Allied military gains. And as occupying power Japan had come to take on some of the features of her Western colonial predecessors. By 1944 doubts about the role of Japan as liberator of Asia and dissatisfaction with military occupation had created unrest of explosive proportions. By August 1944, for example, there were already several incidents of violence between Burman and Japanese soldiers.[1] Other incidents plagued Japanese forces in Atjeh and other parts of Southeast Asia.

By late 1944 and early 1945 it was apparent to Tokyo Headquarters, to intelligence agencies in the field, and to Southeast Asian nationalists alike that Japan's continued presence in Southeast Asia was most precarious. Area army and local army commanders made hasty contingency plans, anticipating massive Allied invasions. In each part of Southeast Asia guerrilla units were trained as a last-minute effort to shore up indigenous defense capability. But among most Japanese-trained indigenous military units there were signs that unrest had built during three and a half years of Japanese occupation to the point where some reaction was imminent. The first revolt of major significance erupted in a *Peta* battalion stationed at Blitar in east Java.

Peta was organized along territorial lines, with two to five *daidan* in each residency or *shū*. There was no central

[1] Kawabe Shōzō, *Biruma nikki shōroku* (excerpts from the Burma diary), unpublished, entries August 1, 3, 24, 28, 1944.

headquarters for *Peta,* no staff corps, and no contact among *daidan,* even within the same residency. This obviated the danger of a nationally organized military revolt against Japanese forces, and in fact restricted the revolt which occurred in Blitar to the *daidan* in that garrison alone.[2]

Deterioration of the war situation for Japan after mid-1943 increased Japanese dependence on Java for materials and manpower. Japan in turn increased demands for greater emergency food production in Java. Farmers had between ten and thirty percent of their harvest requisitioned, often more. The active black market in rice raised the price beyond the purchasing power of hungry farmers. Farmers resorted to eating tapioca and corn to stave off starvation. The slogan "Self-Sacrifice and Serve the Public" was used as an incentive. Japan also sent agricultural technicians to improve quality and yield of rice production.[3]

Mobilization of forced labor was another acute grievance. Called volunteers, the *rōmusha* were mobilized without option. Men between sixteen and forty and unmarried women between sixteen and twenty-five were used.[4] They were used not only in Java but were also sent to Thailand, Burma and Malaya for work on the dreaded Thai-Burma Railway, many of them never to return. Military Administrator General Yamamoto estimates the number of *rōmusha* sent abroad at 140,000-160,000, but some place it as high as 300,000.[5] The forced labor system had a disastrous effect on village life, depriving the village economy of the most vigorous age group. Removal of large segments of the population from the villages had later political and social ramifications as well. It also increased further the pressures on those left cultivating in Java.

[2] Nugroho Nototusanto, "The Revolt of a *Peta* Battalion in Blitar, February 14, 1945", *Asian Studies,* v.7, April 1969, p.112.

[3] Kurasawa Aiko, "Historical Considerations on the Blitar Revolt—toward the origins of Indonesian Independence", (in Japanese), B.A. thesis, Tokyo University, pp.93-4.

[4] Waseda, *Indonesia ni okeru Nihon gunsei shi no kenkyū,* pp.310-2.

[5] Wertheim, W.F. *Indonesian Society in Transition,* p.270. Yamamoto's estimate appears in Waseda, *Indonesia ni okeru Nihon gunsei shi no kenkyū,* p.311.

The Blitar *daidan,* leaving the city in the last half of 1944, was appalled at the misery of the *rōmusha* conscripted locally. *Rōmusha* and the populace in general were by 1944 suffering from malnutrition and disease, especially malaria and dysentry, for which no medical care was available. The economy showed effects of extreme disruption as labor was siphoned away from rice production and other normal tasks. Impoverishment was a primary cause of discontent, in Blitar as elsewhere.

Another *Peta* grievance was that the *daidanchō* and other Indonesian officers were required to salute the lowest Japanese non-commissioned officers. Slapping of officers and men, a normal form of discipline in the Japanese Army, was offensive to Muslims, as Yanagawa pointed out during all the training programs in Java. These problems were common to the INA and BIA as well and produced an unhealthy psychological climate between Japanese and Southeast Asians. There were also reports of girls being taken away by Japanese troops, in Blitar and elsewhere.[6]

Widespread grievances against the Japanese produced numerous small incidents prior to the Blitar Revolt. None of them was organized on a wide scale, and none produced any cumulative energy which might have fed into a war of independence. Part of the reason for this lay in Japan's independence policy. Continuous Japanese promises served to sustain a certain level of popular expectation that Japan would grant independence at some future though never specified date. Japanese policy was successful in defusing any revolutionary potential underlying Indonesian malaise.

Besides these grievances of *Peta* and the populace generally there were some features distinctive to the garrison in Blitar which precipitated revolt there. East Java, which includes Blitar city in Kediri State, was the crucible of the PKI and stronghold of the Indonesian Communist Party. East Java also is the home of many nationalist leaders, including Sukarno, whose home was in Blitar itself. When the Blitar *daidan* was first organized there was a flood of volunteers, seeking food, clothing and the security

[6] Nugroho, "The Revolt of a *Peta* Battalion in Blitar", p.113.

of a job.[7]

The Blitar *daidan* with four *chūdan* came into being on 25 December 1943, with the creation of the first group of *daidan*. The Blitar *chūdan* were sent to coastal areas where they worked with *rōmusha* in the last half of 1944. The economic distress of the *rōmusha* plus the arrogance of some Japanese officers and most *Kempeitai* engendered in the Blitar *daidan* a festering hatred of the Japanese. There were, in addition, incidents involving Indonesian girls. Some leaders of the Blitar revolt, particularly Suprijadi, besides cherished a hope of real independence.

Other causes of revolt have also been cited. Kahin, for example, attributes the outbreak at Blitar to anti-Japanese activity by Sjarifuddin's group and others of the Communist Party, operating underground with Dutch funds supplied before the Japanese invasion.[8] Sjarifuddin, a leftist leader of Gerindo and of the largest anti-Japanese underground, was arrested by the Japanese for suspected anti-Japanese activity in early 1943. In February 1944 he was sentenced to death. Because of Sukarno's intervention his sentence was commuted to life, but at the time of the planning of the Blitar revolt he was actually in prison. Whether or not the activities of Sjarifuddin's group or other underground groups actually did contribute to the Blitar revolt, there were nevertheless enough other causes operating to ignite the outbreak.

Instigator of the revolt was Suprijadi, a quiet young *shodanchō* of religious, mystical bent, given to meditating and fasting since childhood. Suprijadi was one of the first *Peta* officers trained in the Tanggerang *Seinen Dōjō*. The other two leaders, Muradi and Suparjono were *shodanchō* from the Bogor *Renseitai*. Suparjono was deputy commander of the Blitar *daidan*.[9] There were also three *bundanchō* involved in the planning: Sudarmo, Sunanto, and Halir Mangkudidjaja.

[7] Kurasawa, "Historical Considerations on the Blitar Revolt", p.83.

[8] Kahin, *Nationalism and Revolution in Indonesia*, p.114.

[9] Kurasawa, "Historical Considerations on the Blitar Revolt", pp.106-7; Yanagawa, *op. cit.*, pp.178-80; Nugroho, "The Revolt of a *Peta* Battalion in Blitar," p.116.

The *shodanchō* and *bundanchō* began secret meetings in September 1944 to prepare for the revolt. Twelve attended the first session, at which Suprijadi discussed Japanese oppression and his plan for revolt. Those at the meeting agreed to work among the people of Blitar and to contact other *daidan* to organize and plan the revolt.[10] At the fourth secret meeting, held on 4 February, Suprijadi insisted on immediate action because the *Kempeitai* were already aware of the plan. What Suprijadi sought was not simply military resistance but total revolution. Again on 9 February Suprijadi stressed the urgency of immediate action, but the others objected because other *daidan* had not been contacted. A Muslim teacher who acted as confidant of the soldiers advised that the timing was premature. Suprijadi nevertheless urged revolt on the *daidanchō* as well, citing as his reasons the plight of civilians and *rōmusha,* Japanese arrogance, and the need for real independence. The *daidanchō,* however, did not concur with Suprijadi's plan.[11]

Sukarno asserts in his autobiography that he knew beforehand of the plans of the Blitar *daidan.* He was visiting his parents in Blitar when he was met by Suprijadi, informed of the plan, and asked for his opinion. Sukarno allegedly warned Suprijadi that he would have to be prepared to face execution should the revolt fail, and that Sukarno himself would not only not be able to protect the Blitar soldiers but would be forced to disclaim any knowledge of the affair in order to defend the future of *Peta.* "*Peta* is the vital tool in our forthcoming Revolution. I cannot sacrifice my whole army for the sake of a few," Sukarno allegedly warned Suprijadi.[12]

At a midnight meeting on 13 February, Suprijadi finally convinced the others that the Japanese knowledge of their plan made it imperative to act immediately rather than to

[10] Nugroho, "The Revolt of a *Peta* Battalion in Blitar", p.116; Kurasawa, "Historical Considerations on the Blitar Revolt", pp.108-10.

[11] Kurasawa, p.119 quoting Nugroho Nototusanto, *Pemberontakan Tentara Peta Blitar Melawan Djipang,* pp.24-5.

[12] *Sukarno, an Autobiography, as told to Cindy Adams,* p.191. This account is not corroborated by Nugroho or Kurasawa.

sit quietly awaiting arrest. Suprijadi told the men that the Japanese planned to disperse the *daidan* and send all the men out of the city. The *shodanchō* and *bundanchō* were convinced. The revolt began at 3:00 a.m. on 14 February with the firing of mortars at the Sakura Hotel where Japanese officers in Blitar were quartered. The *Kempeitai* headquarters was also hit with machine-gun fire. Both buildings had been vacated, indicating Japanese foreknowledge of the plan. Suprijadi and Muradi gave their soldiers the option of joining the revolt or not. Those who joined were ordered not to fire on any of their fellow-men but to kill all Japanese. A total of 360 officers and men participated in the revolt.[13]

One of the *bundanchō* ordered his men to take down a poster captioned "Indonesia will be free!" and replace it with the slogan "Indonesia is free!" This symbolized the role of Blitar as harbinger of the war of independence. In the course of the revolt some twenty-five Japanese officers were shot.

The revolt was suppressed by a combination of conciliation and force. The *I-go Kimmutai* or First Task Force was deployed against units of the Blitar *daidan,* according to Yanagawa. Through activities of Yoda and Shimizu of the Army Propaganda Section, a meeting was also arranged between one of the revolt leaders, Muradi, and Colonel Katagiri. Muradi presented the following conditions for returning to the *daidan:* 1) participants in the revolt would not be disarmed, and 2) responsibility for the revolt would not be questioned. Some sources add further conditions. As a sign of Colonel Katagiri's acceptance of the terms, he gave his sword to Muradi. The revolt was over and the troops were returned to the *daidan.*[14]

Japan chose to settle the Blitar revolt by negotiation and threat of force rather than by force alone for fear that an attempt at forcible suppression might lead to a widening of the revolt. For this reason, senior officers of the Blitar

[13] Kurasawa, "Historical Considerations on the Blitar Revolt", pp.136-8; Nugroho, "The Revolt of a *Peta* Battalion in Blitar", p.117.

[14] Kurasawa, "Historical Considerations on the Blitar Revolt", pp.156-7.

daidan were called on to use persuasion with their subordinates, and most of the participants were peaceably returned to their units.[15]

Despite Colonel Katagiri's promises, the 16th Army ordered the *Kempeitai* to investigate the revolt, and the leaders were imprisoned and court-martialled. This interrogation was in fact a breach of the agreement between Muradi and Colonel Katagiri that the participants in the revolt would not be interrogated in an attempt to fix responsibility. On the basis of the interrogation seventy-eight officers and men were sent to Djakarta to be court-martialled by the Japanese Army. Six death sentences were imposed and six life imprisonment sentences, with other shorter sentences.[16]

The name of Suprijadi was conspicuously missing from the roster of those court-martialled, for he had disappeared, and has in fact never reappeared since. Instead, he lives as a "phantom general" in the hearts of the Indonesians. Despite his disappearance Suprijadi was named first Public Peace Minister and Supreme Commander of the National Army of the Republic after the war, and by special Presidential order participants in the Blitar revolt were recognized as "fathers of independence".[17]

Despite the existence of both anti-Japanese nationalist sentiment and an armed Indonesian force, there was no armed cohesive Indonesian resistance to Japan during the occupation. *Daidanchō* and *chūdanchō* both indicated after the war that they were holding discussions on anti-Japanese measures, and some believed that if the Blitar revolt had occurred later it might have assumed more widespread proportions. The fact remains, however, that the Blitar revolt was restricted to members of one *daidan*. There was also at least one pro-Dutch *shodanchō* in *Peta* (not at Blitar) who was maintaining contact with other pro-Dutch

[15] *Ibid.*, p.170.
[16] *Ibid.*, pp.172-6; according to Nugroho fifty-five were court-martialled: "The Revolt of a *Peta* Battalion in Blitar", p.121.
[17] Kurasawa, *op. cit.*, pp.180-1, 82, 203-4.

groups while undergoing training in Bogor.[18] But these activities never eventuated in any positive steps against the Japanese occupying power. There were, to be sure, other scattered outbreaks of violence, but none on the scale of the Blitar revolt. Criticism of Japanese occupation never led to resistance of the magnitude that appeared in Burma, Malaya, the Philippines or Thailand.[19]

If hatred of the Japanese was so widespread among the populace generally, how do we explain the absence of nation-wide armed resistance? One fact is the effectiveness of Japanese propaganda. Java heralded the advent of the Japanese Army as liberator of Asians from Western imperialism, and the birth of "Asia for Asians". Military administration in Java and IGHQ in Tokyo were careful not to crush Indonesian aspirations but always to hold out some ray of hope, however dim. The Koiso declaration of September 1944 (whose credibility Suprijadi did not ascribe to) was the last in a series of attempts to stave off popular revolt by holding out yet another prospect, for the first time phrased in terms of independence. The cooperation of Sukarno and Hatta with Japanese Military Administration was also another factor which disinclined the populace toward outright revolution.[20] Another reason for the absence of a generalized *Peta* revolt lay in the military command structure of *Peta*. *Daidan* were commanded separately and did not communicate with each other. Separation of *daidan* was designed specifically to prevent the spread of malaise.

Blitar, nevertheless, from the Indonesian viewpoint represents the beginning of the Indonesian revolution, the 1945 war of independence. Members of other *daidan,* when they heard of the Blitar revolt, reported it influenced their thinking. The Rengasderngklok *daidan,* in particular, feeding directly into the 1945 revolution, was heartened by news of Blitar. The revolutionary energy engendered at Blitar was

[18] Interviews with Effendy Pandjipurnama and Sjachra, ex-*shodanchō* in the First Task Force, Bandung, 30 January 1971; and Kasman Singodimedjo in Djakarta, January 1971.

[19] Waseda, *Studies in Japanese Military Administration in Indonesia,* p.447.

[20] See discussion by Kurasawa, *op. cit.,* pp.100-2.

transmitted to Rengasdengklok, where the soldiers of Rengasdengklok *daidan* kidnapped Sukarno and Hatta.[21] Sukarno and Hatta were taken to a Djakarta suburb and faced with the demand that they proclaim independence immediately, on the night of 16 August 1945. *Daidan* officers also arrested all Japanese in the area, including Yanagawa (by error, according to Yanagawa). The Rengasdengklok incident thus helped precipitate the proclamation of Indonesian independence.[22] And the war of independence was inspired by the Blitar revolt and by the example of Suprijadi, legendary first commander of the National Army.

There were other participants in the kidnapping of Sukarno and Hatta at Rengasdengklok. *Peta* and the Pemudas were supported by leading nationalists, including Sjahrir, Saleh, Wikana, Sukarni, and Dr. Muwardi. In June of 1945 when the Pemudas had decided that the Japanese were the main obstacle to independence and were calling for a struggle against them, Sukarno was still proclaiming his faith in the Japanese side and the obligation to support it. When some outlying areas were already occupied by Allied forces, there was a deep schism between Sukarno and the Pemudas. Sukarno directed his efforts, as he had in the twenties, to forming a united popular movement. The Japanese on 7 August announced a Committee for the Preparation of Indonesian Independence. Sukarno responded publicly, "Dai Nippon has already begun to recognize our independence," and hence there was no reason to rebel against the Japanese. Sjahrir, however, insisted on the need to revolt against the Japanese, and used his persuasive powers to bring Hatta over to his point of view. The Pemudas, angry and impatient, were behind Sjahrir.

Sukarno promised to proclaim independence on the night of 15 August, then announced a postponement until the following day. This procrastination by Sukarno provoked

[21] See discussion in Nugroho, *op. cit.*, pp.56-7; and Kurasawa, "*Peta* and the 1945 Revolution", (in Japanese), ch.1, M.A. thesis, Tokyo University. Also interview with Mrs. Shiraishi in Tokyo, 3 March 1972.

[22] Kurasawa, "Historical Considerations on the Blitar Revolt", pp.194-204.

the Pemudas and *Peta* to kidnap Sukarno and Hatta, and the famous incident occurred at the *Peta* barracks. Sukarno and Hatta were, however, returned, and at the home of Admiral Maeda the same evening the Indonesian declaration of independence was composed. Maeda had worked to prepare Indonesians for independence since the Tokyo announcement of September 1944. The final version of the proclamation, signed by Sukarno and Hatta and providing for the transfer of power, was announced the morning of 17 August. Sukarno had retained his undisputed leadership to the end, walking the difficult path between Japanese control on the one hand and the angry Pemudas and implacable nationalists like Sjahrir and Saleh on the other.

So successful were Japanese political propaganda methods in *Peta* that the *Peta* officers doubted the credibility of the Japanese surrender news in August. Some nationalist leaders attempted unsuccessfully to persuade *Peta* officers either to revolt or to run away with their weapons.[23] *Peta* was successfully disbanded by the Japanese and most of its weapons were in Japanese hands at the cessation of hostilities, though there were some skirmishes.

During 1945 *Peta* fed into the *BKR* (*Badan Keamanan Rakjat,* or People's Security Agency), which in turn fed into the *TKR* (*Tentara Keamanan Rakjat,* or People's Security Army). The *TKR* became the regular army between January 1945 and June 1947.[24] The sequence from *Peta* to *BKR* to *TKR* was significant for army officer groups. As late as 1968 and even 1971 three-fourths of the top officer corps of the Indonesian Army originated with *Peta.*[25]

The political and military confusion of late 1945 was reflected in a phase of disorganized waiting for leadership within *Peta.* Some of the younger officers began to meet to make plans for an Indonesian Army. They were joined by ex-*Seinendan* members, ex-*heiho,* and others in forming

[23] Smail, John, *Bandung in the Early Revolution, 1945-1946, A Study in the Social History of the Indonesian Revolution,* pp.44-5.
[24] *Ibid.,* p.83.
[25] Nugroho, Nototusanto, "The *Peta* Army in Indonesia", mimeo., p.12.

BKR units. In Bandung the *BKR* was born at a meeting of *Peta* officers addressed by Kasman Singodimedjo and other battalion commanders who were assigned the task of forming *BKR* branches. The government in assigning battalion commanders this task sought to perpetuate the role of political liaison they had performed in *Peta* under the Japanese. The principal function of the short-lived transitional *BKR* was to preserve order within the community. During this disturbed period there was some looting, seizing of guns from the Japanese, and even killing of Japanese civilians. Ordinary crowds resorted to force to take control of government offices from the Japanese. In some areas the Japanese resisted and fighting erupted; in others the Japanese relented peaceably. Most weapons were still in Japanese hands at the inception of the revolution. The seizure of control inaugurated the revolution,[26] which continued for several years against the returning Dutch.

In Sumatra the decentralized local organization and lack of a central training center of command of the *Giyūgun,* plus the fact that *Giyūgun* units were under Japanese commanders, meant that there was not only not much *esprit de corps,* but the Sumatra *Giyūgun* could not unite to revolt against its Japanese mentors. There were many incidents between Japanese and Sumatrans, but most of them had no direct connection with the *Giyūgun.* Sumatrans found particularly offensive the Japanese disciplinary technique of slapping. There were several *Pusa*-instigated incidents which resulted in killing of Japanese. Anti-Japanese incidents erupted in Atjeh State as early as 1942. On 10 November 1942 an anti-Japanese riot by armed Atjehnese broke out. To suppress it the Japanese shot over 116 Atjehnese and arrested several more.[27] These outbreaks, particularly in Atjeh, were traceable to the highly explosive religious nationalism of the Atjehnese, which for so many years had been directed at the Dutch. Though the Atjehnese had cooperated with the *F Kikan* during the Japanese landing in Sumatra, this cooperation was predicated on hatred of Dutch rule rather than on

[26] Smail, *Bandung in the Early Revolution,* pp.46-54.
[27] *Tokugawa shiryō* (Tokugawa materials), no.24, unpublished.

acquiescence in Japanese occupation policy, a policy which was not yet apparent. It did not take long for hostility to be transferred from the Dutch to the Japanese once they appeared as the source of oppression, whether religious or economic.

But such incidents were not confined to Atjeh. In July some Batak people attacked a police station south of Medan. Several Sumatrans were killed and injured.[28] None of the anti-Japanese riots was organized on a large scale and none broke out within the *Giyūgun.*

In Burma as elsewhere in Southeast Asia there were also numerous causes of discontent with Japanese military administration. As in Java, these causes resulted in revolt in early 1945 by Japanese-trained forces. But in Burma the revolt was much more widespread, infecting the whole of the BNA. Deterioration of material conditions contributed to malaise of the populace generally, and as in other parts of occupied Southeast Asia fed anti-Japanese sentiment. Requisitioning of materials and mobilization of labor caused critical disruption of the economy and was a festering sore in Japanese-Burmese relations. Air raids and epidemics also contributed to the unrest of the people.[29]

There were other deep-rooted causes of discontent. Among the Thakins and the BNA dissatisfaction with Ba Maw's leadership also lingered. It was nearly impossible, for example, for Ba Maw and Aung San to agree on anything. There was furthermore the often arrogant, insulting behavior of Japanese soldiers, a sensitive point in all of occupied Southeast Asia.

Another factor which heightened nationalistic sentiment was the discrepancy between the proclamation of Burmese "independence" and the reality. On the first anniversary of independence Aung San spoke frankly at the ceremony. "Our independence exists only on paper and the peoples have yet to enjoy its benefits. Only the privileged few such as ministers – like myself – and their hangers-on, the commercial exploiters and the newly-rich reap the fruits of

[28] *Tokugawa shiryō*, no.23.

[29] Interview with Takahashi Hachirō, 15 September 1970, Tokyo; Ōta Tsunezō, *Biruma ni okeru Nihon gunsei shi no kenkyu*, p.444.

this independence. A long hard road still lies between us and our goals," he said.[30]

"Independent Burma" planned to send ambassadors to Thailand, to Manchukuo, and to Nanking, the Japanese-sponsored régimes which recognized Burma, but they never went. Ba Maw also declared that Indians resident in Burma who were British subjects could not be regarded as enemy aliens, and this declaration was reiterated in a court decision and accepted by Japanese authorities. Japanese-supported independence of Burma was after the war held invalid and a contravention of international law. Burma, it was judged, was not a state in international law, but still a British dependency.[31] Basically the same legal arguments were put forth as those advanced in the Red Fort trials regarding the validity of the independence of the FIPG and the INA. "Victors' justice" was reflected in the judgement, as in the decisions of most post-war international military tribunals.

Those who narrate the history of independence movements, in South and Southeast Asia or elsewhere, have been anxious to establish the antiquity of the origins of their struggles. U Ba Than is no exception. He dates anti-Japanese resistance in the BIA from as early as the training of the original thirty in Hainan. Ba Than describes Aung San's meeting with the "underground" at the time of his return to Burma in March 1941. Those at the meeting, Ba Than writes, were given to understand that Japanese troops would not enter Burma. Trained Burmans would enter Burma, Ba Than maintains, but Japanese troops would not. During Hainan map maneuvers, however, map exercises posited BIA and Japanese units fighting together. These exercises made the thirty apprehensive about Japan's plans, alleges Ba Than. Besides Japan's failure to keep promises regarding independence and the use of Japanese troops in Burma, Ba Than cited *Kempeitai* torture tactics and arrests as factors contributing to Burma's disillusion with Japan.[32]

[30] Quoted from the daily newspaper *New Light of Burma*, 6 August 1944 by Maung Maung, in *Burma in the Family of Nations*, pp.98-101; Cady, John, *A History of Modern Burma*, p.480.

[31] Maung Maung, *Burma in the Family of Nations*, p.101.

[32] Ba Than, *The Roots of Revolution*, pp.42-3.

Kempeitai intelligence reports also mentioned preparations for revolt at least by 1943, but most of the Burma Area Army command was not convinced of the seriousness of the problem. When in 1945 advisers Sawamoto and then Sakurai concocted last-minute plans to stave off revolt, their schemes were not approved by their superiors, and in any case came too late.

There were other causes of dissatisfaction with Japan among BIA officers. One early cause was Japanese occupation of the Governor-General's residence and use of this residence as headquarters of Military Administration. The BIA had anticipated use of the Governor-General's residence symbolically as its own headquarters in Rangoon. That Japan instead used it not for the BIA but as Military Administration Headquarters was taken as symbolic of the real relationship between the Japanese Army and the BIA and Burma. This caused bitterness among the BIA officer corps.[33]

Some officers also saw the reorganization of the BIA as the BDA as a Japanese attempt to reassert more direct control over the BIA, which in a sense it was. Nineteen of the younger officers accordingly formed a resistance group which in October 1942 began distributing anti-Japanese leaflets.[34] In secret meetings the young officers discussed issues of ideology, strategy, tactics, organization and operational orders. The underground group also had a song in which the Japanese were described in pejorative terms.[35] Leading conspirators of the resistance were Bo Aung Gyi, Bo Tin Pe, Bo Chit Khine, Bo Aye Maung, and Bo Ye Htut. Bo Khine Maung Gale, Bo Win and Bo Maung Maung used the office of Colonel Bo Ne Win as headquarters for their activities.[36] Organizationally the conspirators were aided in having a unified Burman command, an advantage which Javans in *Peta* did not enjoy.

The effect of these plans was to precipitate formation of a Planning Staff in the War Ministry. The Imphal disaster

[33] *Ibid.*, pp.31-2.
[34] Ōta Tsunezō, *Biruma ni okeru Nihon gunsei shi no kenkyū*, p.443.
[35] Ba Than, *The Roots of Revolution*, p.44.
[36] Ōta Tsunezō, *Biruma ni okeru Nihon gunsei shi no kenkyū*, p.443.

added another dimension to BDA underground planning. By March 1944, in the midst of the Imphal campaign, the mood of some officers became increasingly insistent. An added irritant was the fact that the Japanese Army had virtually ignored the BDA during the Imphal campaign. They fixed 22 June as the day of revolt, while the Japanese Army was bogged down at Imphal. Ne Win reported the plan to Aung San in Rangoon. Aung San, however, felt the timing was precipitous and unrealistic. He objected to the concept of revolt against both Japanese and British simultaneously and favored preliminary negotiations with the British Army and a British guarantee of complete independence as a precondition of revolt. The resistance group plan had called for intercepting communication routes and attacking both British and Japanese troops.[37] As a result of discussions between Ne Win and Aung San the order was postponed to 22 July. The young officers, impatient at the caution and vacillation of commanders, contemplated kidnapping Aung San if he continued to oppose an immediate rising.

But Aung San was affected by the impatience of the resistance group. Speaking at the August 1944 rally on the anniversary of Burmese independence he said, "What is freedom, and where is it? The truth is that the freedom we have in Burma today is only on paper. It will be a long time before it is a reality." Ba Maw was displeased with Aung San's words, which elicited applause from the audience.[38]

Resistance was stepped up in August in Lower Burma through the actions of the young officers, who won over Thakin Soe, Communist Party leader, to the plan for revolt. The young officers formed a group called "Vanguard of the Revolution". With the formation in September 1944 of the forerunner of the Anti-Fascist People's Freedom League, plans for revolt became more widespread and open.[39] The Army printed the League's manifesto and distributed it throughout Burma. It exhorted all to cooperate with the

[37] Ba Than, *The Roots of Revolution,* p.47.
[38] *Ibid.,* p.40; Ota Tsunezó, *Biruma ni okeru Nihon gunsei shi no kenkyū,* p.444.
[39] Ba Than, *The Roots of Revolution,* p.49.

League, prophesied disaster for Japanese fascists and British imperialists, and proclaimed a program of action to eliminate these evils. Resistance was organized in eight areas, each to be under a military commander and political commissar. The All-Burma Youth League and political groups among the Karens, Shans, and Kachins also joined the AFPFL, which took on a national character.[40]

Aung San was now committed to revolt and involved in the preparations. He read a public resolution at the September 1944 meeting of the AFPFL calling for driving the "fascist Japanese" from Burma.[41]

Tensions between upper levels of the BDA command and the young resistance leaders over the timing of the revolt in fact led to postponement of action until March 1945. But guerrilla acts against the Japanese began in Arakan in November 1944, spilling over into Akyab. Some officers felt Communist leader Thakin Soe was temporizing. At a secret meeting at Mingaladon headquarters in the 4th Battalion they fixed D-Day for November. Again Army authorities intervened and postponed the date to 1945.

British-sponsored Force 136, which was infiltrating occupied Malaya and Thailand during the war, also made contact with the resistance and the AFPFL guerrillas in Arakan, distributing British weapons. 3,000 arms were dropped by Force 136 or issued to resistance units and the Communist Party, a British tactic also used by Force 136 in Malaya. At the same time 12,000 arms were issued to Karen units, reflecting suspicions which lingered in the minds of British military intelligence regarding the reliability of the BNA.[42]

The Japanese Military Administration was naturally concerned about reports of disaffection from Japan within the BNA. Adviser Sawamoto initiated conversations with Aung San on 24 September 1944 and was reassured temporarily by the results. He recorded in his diary: "The atmosphere of anti-Japanese resistance which I sometimes have felt

[40] *Ibid.*, p.50; Ōta Tsunezō, *Biruma ni okeru Nihon gunsei shi no kenkyū*, p.445; Cady, *A History of Modern Burma*, p.465.
[41] Izumiya, *Minami Kikan*, p.222.
[42] Ba Than, *The Roots of Revolution*, p.53.

since the beginning of last month is nothing to worry about if the general doesn't change his mind."[43]

Sawamoto proposed an expansion in the size of the BNA to further offset the danger of dissatisfaction within BNA ranks. The Sawamoto proposal was approved by BAA commander General Kimura in Maimyo. By this date, however, confusion was general within the Japanese Army command in the aftermath of the Imphal debacle. There was furthermore a serious problem of weapons shortage. Availability of weapons placed a limit on possibilities of further expansion of the BNA. This was already a problem in mid-1944, when arms of the BNA were compared unfavorably with those of the INA. Further plans for expansion of the BNA to as large as 300,000, drafted in March 1945, were beyond the realm of possibility and already too late to alleviate BNA tensions.[44]

Sawamoto met Aung San for a second conversation in mid-February 1945. Aung San was obviously shaken by reports when he asked Sawamoto, "I have heard the British Army has crossed the Irrawaddy. What is the Japanese Army going to do now?" Sawamoto explained Japanese counter-plans as best he could. Shortly afterwards Sawamoto was transferred to command of the 33rd Division in north Burma and regretted having to leave the BNA. Sawamoto's replacement, Major-General Sakurai Tokutarō, immediately planned a spectacular expansion in the size of the BNA, which was never carried out.[45]

The military situation deteriorated rapidly for Japan after the fiasco at Imphal. The British counter-offensive by land began with the occupation of a beachhead on the coast of Singapore early in January 1945. British forces pushed into Burma, and the Japanese Army was unable to prevent the crossing of the Chindwin or the Irrawaddy. British-Indian forces overcame Meiktila; Mandalay in Upper Burma fell to them on 18 March. Simultaneous attacks were mounted from East and West on the Arakan mountain region and by sea as well. Japan's precarious position in

[43] Sawamoto, *Nihon de mita Biruma gun no seiritsu*, III, pp.406-7.
[44] *Ibid.*, pp.407-8, 417-19.
[45] *Ibid.*, pp.417-18.

Burma was apparent to all. It provided the opportunity the BNA and AFPFL had been awaiting.

The earliest outbreak of revolt against the Japanese was a declaration of war against the Japanese by Bo Ba Htut, BNA commander at Mandalay on 8 March, followed by open action against the Japanese with several hundred men. Ba Htut and his battalion crossed sides and joined the Allied invading forces. Everywhere quarrels and incidents were erupting between Japanese and Burmans.

As an emergency measure the Japanese Army dissolved its Advisory Department and organized a *tokumu kikan* designed to prevent generalized revolt. It failed in its mission, for the causes of disaffection had already reached mammoth proportions.[46] General Sakurai discussed the Bo Ba Htut rising with other advisers. The problem was whether the BNA could be trusted at all or should be disarmed. The conclusion reached was that the Bo Ba Htut revolt was an isolated instance, that Bo Ba Htut had acted on his own, and that the revolt was therefore not typical of the whole BNA. There was therefore no need of disarming the BNA. Rather, more arms should be issued to the BNA and more trust shown by the Japanese Army. It was a fateful miscalculation.

Sakurai also met with BNA officers to plan joint action with the Japanese Army following British crossing of the Irrawaddy River. On 16 March 1945 a Defense Council of Japanese and Burman officers was held at the Governor-General's residence. BAA commander General Kimura, Chief of Staff Tanaka, and Sakurai met with Aung San, Ba Maw, Ne Win and Bo Ze Ya. Plans for departure of the BNA to the front were formalized. The following day a review of battle units was held. Ba Maw was notable among speakers at this military review.[47] Departure of the main force of the BNA, scheduled for 20 March, was postponed until 24 March. This was another serious Japanese miscalculation. Continuous Allied air attacks contributed also to increasing unrest among BNA units poised for action

[46] Interview with Takahashi and Kawashima, 3 July 1972, Tokyo.
[47] Sawamoto, *Nihon de mita Biruma gun no seiritsu*, III, p.420.

with Japanese forces. By 24 March there were already reports that units in Pegu and Mandalay had made contact with the British.

Officer candidates at Mingaladon held an emergency meeting on 23 March. After the meeting the cadets went through the gates of the academy, never to return.

Sakurai, with Captain Takahashi, who had been with the BIA since its inception, were with Chief of Staff Bo Ze Ya at Shwedaung on 26 March. On 27 March Aung San disappeared from Shwedaung, and Sakurai and other officers were unable to find him.

Sakurai, attempting to salvage some remnant of co-operation and harmony, declared the revolt was directed against Ba Maw rather than against the Japanese. He made a last-ditch proposal for the restoration of the Burmese monarchy, similar to the Japanese support of Bao Dai in Indochina. On 29 March Yan Naing, commandant of the Mingaladon officer school, returned to Rangoon. Sakurai hoped to form a new army around Yan Naing. The Japanese Army, however, failed to approve either Sakurai's suggestion for a revived monarchy or for an enlarged and reorganized army, and Sakurai too was transferred, returning to Japan on 1 April.[49]

When revolt broke out, several Japanese officers were killed by BNA troops. But not all Japanese officers were targets of the revolt and not all BNA officers joined the action. Takahashi was protected by Ne Win. And one of the original thirty, Bo Mying Aung, killed himself rather than turn against his Japanese teachers.[50]

The BNA continued to fight from 27 March until May 1945, when Aung San met with General Slim and later with Mountbatten in Delhi. On 30 May the BNA was recognized by the British as the PBF, the Patriotic Burmese Forces. This was followed by a victory parade in Rangoon on 15 June. The PBF became the first issue over which British and Burmese leadership negotiated at the end of

[48] *Ibid.*, pp.423-4.

[49] Interview with General Sakurai, 10 December 1970, Fukuoka City; Sawamoto, *Nihon de mita Biruma gun no seiritsu*, III, pp.434-5.

[50] Izumiya, *Minami Kikan*, p.225.

hostilities.

In May 1945 a British White Paper was issued which envisaged the return of direct rule for three years, to be followed by the election of a legislature and Burmese Council, with restoration of the Act of 1935. On 15 June 1945 Aung San, meeting with Mountbatten, promised to cooperate and agreed that the BNA be absorbed into the regular Burma Army, for those who volunteered to join. Those who did not volunteer would be disbanded. When British Governor-General Dorman-Smith returned to Burma at the end of the war he promised in a public speech to repair the damage of war and to see full self-government established in the shortest possible time. This was unacceptable to the AFPFL, which did not wish to cooperate in a restoration of British rule. Aung San wrote Mountbatten that he would have to resign from the Army, as the AFPFL could not agree to the resumption of British rule. Mountbatten then disbanded the BNA, announcing at the same time formation of a Burma Army. He offered the Deputy Inspector Generalship to Aung San who refused, saying he was giving up his military career to enter politics. Dorman-Smith's attempts to win over the AFPFL were unavailing.

Aung San then founded the People's Volunteer Organization with officers and men from the BNA who had opted not to volunteer for the Burma Army. The BNA was thus re-born. Aung San at the same time informed the British that it was not an army but a Home Guard. The PVO thus transmitted the spirit, organization, and experience of the BIA to the post-independence army of Burma.[51]

From the Japanese standpoint the policy of not appointing a Javanese supreme commander of *Peta* paid off in 1945. Since there was no official communication among *daidan* and no commander-in-chief of *Peta,* the Blitar revolt was confined to the Blitar *daidan,* and to only part of that *daidan.*

In Burma, on the other hand, the BNA had a Burmese Commander-in-Chief and Chief of Staff. The plan for revolt was discussed in staff councils. Aung San and Ne Win were both consulted by the conspirators, and disagreement

[51] Collis, Maurice, *Last and First in Burma,* pp.244-64.

among top-ranking staff officers over the timing of the revolt in fact led to its postponment. But the BNA revolt, when it occurred, involved nearly the whole of the BNA. 16th Army staff officers in Java, comparing the Java situation with Burma, were justified in judging the *Peta* situation more stable than the BNA.

The INA, devoted as it was to the absolute leadership of Bose, never revolted against the Japanese, though there were some incidents between soldiers. Bose never officially turned his back on the Japanese or contemplated negotiations with the returning British. He received Japanese aid in his end-of-the-war plan to go to Russia for help against the British. The INA thus shared the military fate of Japan, and INA officers have assumed roles of political rather than military significance in independent India.

Though the Japanese in organizing and training these armies presupposed Japanese control over their actions, this premise proved false in the end. Through revolts against their Japanese training officers, *Peta* and the BNA both demonstrated their character as genuine independence armies. These armies in revolt led directly into the armies of post-war Burma and Indonesia, in Java through the war of independence and in Burma through negotiations with the British leading to independence and through the PVO.

7

The Significance of the Japanese Military Model for Southeast Asia

The impact of the Japanese training of these independence and volunteer armies on Southeast Asia went beyond the original Japanese motives in organizing and training them. Consider the total impact of this military imprinting on Southeast Asia. What did Japanese military training come to mean for Asians in post-war independent Southeast Asia? The answer to this question is different for each nation and for each army. One dimension of the issue relates to the military role of the indigenous armies during the war. Another aspect relates to the impact of Japan's collapse on the fate of independence movements. Another dimension concerns the individual post-war career profiles of the Japanese-trained officers and men. Yet another aspect of the question relates to the collective patterns of integration of these men into post-war military and/or political élites of the nations of post-war Southeast Asia.

What became a potent generator of internal forces for revolution, apart from the military training itself, was that the Japanese selected for special education and training especially in Burma and Indonesia segments of potential leadership which had been excluded by Western colonial régimes. In Burma, for example, political leaders imprisoned by the British, including Ne Win and Ba Maw, were released. For military training, the Japanese selected as the original "thirty comrades" young nationalist students of the Thakin Party. Already politically aware, and in some cases in prison or in exile, these men were stimulated politically and given military expertise by their Japanese tutors. Similarly, in the BIA the Japanese recruited mostly Burmans, avoiding Chins, Kachins, Karens and Shans who had served in the colonial forces. By shunning groups which had served under the

British and encouraging groups which had not, the Japanese occupation injected potent forces for social change into the Burmese political and military scene.

Similar policies in Java and Sumatra dictated choosing nationalist leaders who had been imprisoned or ignored by the Dutch. The Japanese deliberately avoided recruiting for military training or political positions groups such as the Ambonese, Menadonese and *ulèëbalang,* who had been recruited into the *KNIL* and the Dutch colonial bureaucracy. Instead the Japanese selected young men recommended in many cases by powerful Muslim leaders or by local village officials. And as in Burma, imprisoned leaders like Sukarno were released and encouraged politically. The result was, as in the Burmese case, the rise of previously disfranchised groups to the top politically and militarily.

The INA, one of the most important armies considered here, was the only one to fight beside the Japanese Army in a major campaign. There were several reasons for Japan's including the INA in the Imphal offensive, while refraining, for example, from using the BIA in the same campaigns in Burma or Imphal. The most important difference from a strictly military standpoint was the superiority of the INA as a battle-ready fighting force. Most of the officers and men of the INA were already trained, seasoned soldiers with long experience in the tradition of the British Indian Army. Japanese encouragement of the INA was therefore a political, liaison function more than basic military training for military motives or a part of a military administration.

The BIA, on the other hand – and similarly *Peta* and all the Japanese-trained *Giyūgun* as well – was a force recruited, organized and trained from scratch, without British or Dutch antecedents. Though the original Burmese "thirty" were organized even before the Pacific War, in 1944 Japanese officers still harbored serious doubts about the battle-ready character of the BIA. "We felt they were not so sharp as a fighting force," one Japanese officer close to the BIA during the entire war recalled.

The formation of the *Peta* and the *Giyūgun* in Malaya, Sumatra and Indochina, however, took place later than the

INA or BIA, after Japanese occupation of Southeast Asia. They were organized for limited military objectives, primarily to supplement deficient Japanese defensive strength in anticipation of renewed Allied counter-offensives. The exceptions to this statement were *Peta* and *Makapili,* where Japanese aims were also partly political owing to the strength of Indonesian and Filipino nationalism.

None of the forces other than the INA had a highly trained staff with whom Japanese officers felt they could conduct joint strategic planning sessions and map maneuvers. Even with the INA during planning for the Imphal offensive, Japanese staff officers harbored doubts about the caliber of the INA as a fighting force and did not include INA officers in serious strategic decision making.

For post-war Southeast Asia, however, the volunteer and independence armies assumed a significance ultimately greater than Japan foresaw. They fed directly into the armies of newly independent Southeast Asian nations, contributing to independence wars and armies of the post-war era. They furnished highly trained officers who at least in Burma and Indonesia could perform staff functions. The significance of the Japanese model for Southeast Asian staff officers was further accentuated through the training of selected Burmans, Indians, Thais, Indonesians, Filipinos and Malays (as well as Koreans and Manchurians) at the Military Academy in Japan (and in Manchukuo). The Japanese military model has therefore been far more influential for Burma and Indonesia and other areas than for India's army. As one Indonesian general put it, "The Japanese taught us something far more important [than the Dutch]: how to create an army from scratch."

Another dimension of the impact of the Japanese military model was the emphasis on training guerrilla units. Guerrilla training was a universal feature of Japanese military training programs in Southeast Asia by contrast with the lack of such training by Western colonial powers. In the view of some Indonesian generals this colonial oversight accounted for the brevity of the Dutch resistance to Japanese invasion of Java and Sumatra. Similar analyses have been made by Indians and Malays regarding the Japanese invasion of Malaya and other parts of continental Southeast Asia. The British were not equipped to deal with Japanese guerrilla tactics, despite the much longer British military presence in Southeast Asian jungles. Guerrilla training is therefore a legacy still acknowledged to Japanese military instruction by Southeast Asian officers.

Another aspect of the Japanese military contribution to post-war Southeast Asia is that many Japanese soldiers remained behind in Southeast Asia after Japan's defeat, in the Indonesian case to join the Indonesian war of independence against the returning Dutch. Estimates are that approximately 350 to 400 Japanese soldiers remained in both Java and Sumatra at the end of the war and joined the war of independence.[1] This direct contribution of individual Japanese is another aspect of Japan's wartime military legacy to Southeast Asia.

Two basic assumptions underlay Japanese military training programs, both in the Japanese Army and in the armies in Southeast Asia. One was that spirit, *seishin,* is more important than any technological advantage in military weaponry. The other was that self-discipline must be absolute, precluding any conflict with other values. One example of a conflict in values occurred in Sumatra, where Japanese instructors of the *Giyūgun* were distressed when young recruits stopped for Muslim prayers in the midst of military maneuvers. In the more rigorous training programs in Java and Burma such nonchalant straying from discipline did not occur. The emphasis on spirit was designed to encourage heroism in battle when all technical military advantages might lie with the enemy. The discipline, endurance and confidence thereby instilled were in all cases transferred to nascent armies in post-war Southeast Asia. In evaluating the legacy of Japanese training, the special emphasis on inculcating a fighting spirit, self-reliance and self-discipline cannot

[1] Some estimates of the number of Japanese who joined the Indonesian war of independence are much higher, as high as 1,000: interview with Kawadji Susumu, July 1972, Tokyo. The author met one of these Japanese expatriates in Bandung, a man who continues to live in Java and who operates an auto repair shop, a not atypical vocation for such ex-soldiers in Indonesia.

be overemphasized. The Japanese imparted to their Southeast Asian trainees the assumption that the élan of the warrior was far superior to his technical expertise and would enable him to overcome any obstacle. Detection of the seeds of this spirit among applicants for Japanese training programs, and fostering the spirit among trainees were mentioned by many Southeast Asians as the foremost feature of Japanese education, both military and civilian. In this respect there was no great distinction between education in wartime Japan and Southeast Asia. Many Southeast Asians note today that they still feel the effects of this spiritual training, though the impact of the technical education was more ephemeral.[2] In Indonesia emphasis on these traits produced what is known as the "*Peta* officer type".

In Burma and Indonesia the recruits were mostly young students who had not previously considered entering a military career. Even after they were recruited and trained they generally did not regard themselves as professionals in sense of the Dutch, British, or British Indian Army career officers. Their commitment was to independence, not to careerism. In this respect there was a kind of spiritual harmony between the Japanese tutors and their Southeast Asian trainees. "We didn't become soldiers for money," said General Bambang Sugeng, former Chief of Staff in the Indonesian Army. He hesitated to use the word "professional"[3] at all. Even with the INA, whose members had been British-trained professionals, they volunteered for the INA and cooperated with the Japanese primarily in the interests of Indian independence.

What has the INA experience meant for independent India, if they did not gain their military training from the Japanese? This assessment is conditioned by the fact that the INA experience, like the other Japanese-trained armies, was so ephemeral. Because of the brevity of the experience some advantages in clarity of perception are offset by the absence of continuities and antecedents. Bose's death in 1945, for example, makes it impossible to project with certainty what

[2] This conclusion is based on many interviews by the author in Thailand, Indonesia, Malaysia, and Singapore, January-February 1971.
[3] Interview 19 January 1971, Djakarta.

might have been the political-military legacy of the INA to India had he lived. This suggests not so much that the significance of the INA was limited to a single individual as that the major goal of the INA was achieved with independence in 1947.

The British military model, more than the Japanese, provided continuities in precedent and tradition for the INA. Military experience was not something new for INA officers, though the self-confidence they gained in an Indianized army was. After the war most INA officers and men were interrogated, tried and dismissed, leaving the pre-war military structure otherwise intact. The Indian Army officer corps retains the British tradition of separation between military and civilian spheres, coupled with disdain of politics by the military.[4]

There is, nonetheless, no problem of demonstrating the usefulness of the Japanese to the INA, for without the Japanese Army the INA could have harbored no real hope for liberating India. But the real significance of the Japanese model for post-war India is closely connected with the metamorphosis of the INA officer class.

Bose's vision of free India was realized. His other mission – the communal harmony so distinctive but so ephemeral in the INA – was also briefly revived in the 1968 attempt to forge a political party out of the vestiges of the INA. The party, named the *Azad Hind Sangh,* never really came to life. Again, the reason for this lack of viability is that the goals of the party were partly anachronistic; the chief aim of the INA had already been realized with independence. The commual harmony and cohesiveness which so distinctively characterized the INA was achieved in part through resistance, first to British then to Japanese control. The result was an INA – by contrast for example with the ethnically and geographically limited BIA or *Peta* – which was in a real sense an all-India army. Symbolically, among the three top INA officers tried at the Red Fort tribunal were a Hindu, a

[4] Shils, Edward, "The Military in the Development of New Nations", pp.39-40 in Johnson, ed., *The Role of the Military in Underdeveloped Countries;* Evans, Humphrey, *Thimayya of India, a Soldier's Life,* pp. 283-4. See also discussion in Cohen, Stephen, *The Indian Army.*

Muslim, and a Sikh. Some allege that, had Bose returned to political leadership in independent India, this transcendence of communal animosities would have been a major legacy of the INA. Perhaps it might even have averted partition. But with the negative external factor of Britain or Japan removed, and given other events which have transpired, this remains a moot point. One of the fascinating aspects of Bose's appeal and leadership was that, despite his personal religiosity, unlike other Indian nationalist leaders (with the possible exception of Nehru and Mrs. Gandhi) he did not use religious symbols for political purposes, possibly out of the realization that they would have divided Hindu from Muslim and Hindu from Sikh.

The INA has not become a cohesive political-military élite replacing pre-war nationalist leadership. Nor has the man on horseback – German or otherwise-inspired – found a real place in the post-war Indian politique. Something else has happened to the INA officer corps. Both professional military men and civilians in the INA and FIPG have been politicized and bureaucratized through their experience. Rather than return to their military careers from which they were purged, they have turned instead to politics. Some INA veterans have been elected to Parliament, others have received high-level diplomatic appointments, and still others have entered the lower echelons of the bureaucracy. Some have also gone into labor union leadership. That this politicization occurred with the INA but not with the regular Indian Army suggests that the Japanese interlude acted as a catalyst in the metamorphosis. This is a kind of inversion of what Silverstein describes as an emphasis on the role of the military through Japanese influence.[5] This inversion may reflect the pre-war Indian military experience by contrast with its absence in pre-war Southeast Asia. It may also demonstrate the thesis that an army in defeat is an army

[5] Silverstein, Josef, "The Importance of the Japanese Occupation of Southeast Asia to the Political Scientist", p.8, in Silverstein, ed., *Southeast Asia in World War II*. In Pakistan, however, some former INA officers have led early post-war military expeditions into Kashmir. Others have remained in the Pakistan Army and survive under Ayub Khan as political officers. This information comes from a conversation with Stephen Cohen.

politicized. But *Peta,* on the contrary, victorious in its war of independence, was also politicized.

The INA experience was revolutionary on more than one level. First, as a direct revolution against British rule the INA was partially successful through the British response to the Indian atmosphere surrounding the court-martial in New Delhi. Second, as an indirect revolution within the context of Japanese cooperation, the officer class was transformed. It was bureaucratized and politicized, and in the process absorbed into the political élite of independent India.

The training given various groups of Burmans by Japan during the war had a dual impact not paralleled in the case of the INA but with a closer analogy to the case of *Peta.* Since the INA officers were for the most part already trained in the British Indian Army, the Japanese experience had for these Indians a basically political impact. They were in fact separated from the army of post-independent India and instead assumed roles of political leadership.

It was also easier for the Japanese Army to use the BIA-BDA for its own purposes than to use the INA. One reason for this difference was the schism in Burmese leadership between the civilian Ba Maw and the Thakins in the BIA-BDA-BNA. In the Indian case, civil and military leadership was combined in Bose, who *ipso facto* had more bargaining power with the Japanese than did Ba Maw or the BDA. Another factor was the prior military training and experience of the INA by contrast with the BIA. For this reason, the INA is at one end of the spectrum on the puppet-to-independence army scale. Japan faced many frustrations in attempting to control the INA. Perhaps the relatively loose Japanese control of the INA helped to prevent an INA revolt against the Japanese, unlike the BNA and *Peta.* INA leadership turned instead to calculating alternative sources of help for the cause of Indian independence at the end of the war. The INA, unlike other armies, was never completely under Japan's control.

With the BIA Japanese training became a two-edged sword. It provided BIA officers initial basic military training and staff officer training, enabling them thereby to create and train the Army of independent Burma. But beyond this it gave the inexperienced but politically aware Thakins political experience and technical expertise which have served them well in the military bureaucracy of post-war Burma. It strengthened their nationalism and gave them military, thus revolutionary, potential. This potential the BNA directed against Japan in an all-out revolt more far-reaching than any action taken by other Japanese-trained armies. The BNA transferred both political and military skills to post-war Burmese leadership, refusing to accede to a return of British rule. These results, though inadvertent and not foreseen by all their Japanese tutors during the war, are the outstanding heritage of wartime Japan in post-war Burma.

What has been the significance of *Peta* for Indonesia? Whether or not an army or political group were puppets is in part a question of degree, in part a matter of motive force and initiative. First, there is a difference in emphasis on the source of initiative for the creation of *Peta* according to the account, as has been noted. Yanagawa and Inada stress the Japanese initiative, while Gatot and Sukarno stress the Indonesian role. Both Nugroho and Kurasawa feel that Gatot was being used by the Japanese, and Kurasawa uses the term "puppet army" to describe *Peta*. There seems to be no quarrel with this conclusion. Another dimension of the question is: were the Indonesians in turn as successful in using Japanese support and training for their own aims as were the INA and BIA? The answer to this question is also negative, at least at the time of Japanese occupation. But this is not to say that the men trained in *Peta* did not benefit in their own terms during the war of independence or in the post-war period from their Japanese experience.

Yanagawa, who was perhaps more closely involved with the training of *Peta* in all its phases than any other Japanese, was criticized by others in the Military Administration at the time of the Blitar revolt, on the ground that he had created an instrument which came to be used against the Japanese. In his own defense he states, "My six children in Bandung became fifty people, then a hundred, a thousand, ten thousand, a hundred thousand. Those who fought in the war of independence scattered as flowers, fought to the last, and won Indonesia *merdeka*. For this achievement I still feel proud. Lastly, I don't feel that the Pacific War was useless. This I want to say. It achieved the great aim of emancipation of the people."[6]

The rationale of *Peta* also represented an innovation: fighting to defend the homeland. Hatta pointed to the significant difference between the pre-war *KNIL,* trained to fight the enemy within, and *Peta,* trained to fight the enemy without.[7] *Peta,* like the BIA, strengthened nationalism and gave it revolutionary potential.

The post-war career profiles of the Japanese-trained Southeast Asian participants are often as significant as their wartime activities, reflecting the INA post-war profiles. Of the original Burman "thirty comrades", one died in training before the war, six were killed or died during the war, one, Aung San, was assassinated while Premier; four were arrested by Ne Win; two resigned or retired from politics; three assumed leading roles in the Communist Party and one in the Socialist Party; a few have died since the war; and three joined U Nu's armed resistance organization on the Thai-Burma border, from which point they planned the overthrow of Ne Win. (U Nu has since abandoned the endeavor and sought political asylum in India).

The Burma Revolutionary Party has been staffed since 1962, then, not by the original "thirty", with the exception of Ne Win, but rather by the "second generation" officers of the BDA phase. These include some officers who studied at Mingaladon, a few who went on in addition to the *Shikan Gakkō* in Japan. As with the INA, many of this group have been given ambassadorial appointments. Some have served also as military attachés in Thailand, Indonesia, Australia, Japan and Washington. An example is Col. Mya Thaung who in 1971 was military attaché in Thailand.

As with the BIA and INA, the *Peta* training under the Japanese acted as catalyst in militarizing and politicizing a whole generation of Indonesian leadership. Ex-*Peta* officers are still the most significant segment of the political-military élite in Indonesia. Some *KNIL* Dutch-trained officers still survive, and to these have been added the officers trained

[6] Yanagawa, "Jawa no Beppan", p.155.
[7] Kanahele, *op. cit.,* p.132.

since 1945, partly under American influence. According to the most authoritative estimate, however, three-fourths of the top ranks in the Army are still (in 1971) staffed with ex-*Peta* officers.[8] These men came particularly from the *chūdanchō* ranks. Some of them have additionally served in ambassadorial posts, which appears to be a common career pattern among the most important Japanese-trained army veterans. Among the seventeen Indonesians sent to the *Shikan Gakkō,* one, Gen. Yoga Sugomo, has the highest position in military intelligence (1971). Another, Omar Tusin, has since the war left the army and become a prominent public figure and businessman trading with Japan. As elsewhere in Southeast Asia, Japan imparted to Javans the will to resist through their own efforts and precipitated incidents which fed into the war of independence.

In post-war Sumatra, as in the rest of Southeast Asia, Japanese-trained officers have risen to prominence in political and military roles. Lt.-Gen. Maraden Panggabean, present (1973) Indonesian Minister of Defense, is a Toba Batak from Tapanuli trained in the *Giyūgun* in that area. Similarly, Maj.-Gen. Ahmad Tahir, presently in the diplomatic corps, was between 1969 and 1972 military commander of all of Sumatra. He was a member of the *Giyūgun* in East Sumatra, rising to the highest Sumatran rank of *shotaichō.* Some other *Giyūgun* members who also rose to prominence in post-war Sumatra lost out in the Outer Islands Rebellion, as did their fellow Sumatran in Java, Col. Lubis. Some of these, like Lubis, have under Suharto again re-entered the military bureaucracy.[9]

Veterans of the volunteer armies in Malaya, Indochina and the Philippines, on the other hand, have not played as prominent post-war roles as the independence army veterans. Those who have become prominent, for example in Malaya, have not been anxious to publicize their wartime Japanese connections. The relative post-war anonymity of this group of Southeast Asians reflects also the limited goals of these units and the relative obscurity of the Japanese who trained

[8] Nugroho, "The *Peta* Army in Indonesia", p.12.
[9] From a conversation with Dr. Anthony Reid of Australian National University.

them. As has been noted, the volunteer armies were in general not trained by *Nakano Gakkō* graduates or even in all cases by commissioned officers. It has been virtually impossible to locate any of the Japanese involved in the training of the Malayan *Giyūgun* or *Giyūtai*. The volunteer armies, therefore, with the possible exception of Sumatra, present a different typology in post-war career profiles as well as in training from the independence armies.

The Japanese who trained the independence armies have in many instances retained their close ties with Southeast Asia and with the men they trained. Former *F Kikan* members maintain the closest and most spectacular post-war ties with their wartime students. One *F Kikan* member, Kunizuka Kazunori, who wrote a book on his wartime experiences, lived for many years after the war in India with connections with Mitsubishi. Another *F Kikan* member, Ishikawa Yoshiaki, is in the Japanese Embassy in Pakistan. And Fujiwara Iwaichi, chief of the wartime *F Kikan,* has since the war commanded the first division of the Ground Self Defense Forces. Since retirement he has campaigned unsuccessfully for election to the House of Councillors. His campaign manager was Izumiya Tatsurō, former member of the *Minami Kikan* and author of a volume on the *Minami Kikan*. In 1973 Fujiwara organized the Fujiwara Asia Research Institute, which includes other veterans of the *Kikan* as well. Fujiwara's ties with INA veterans are particularly close and warm. He has encouraged his nephew to study seven Asian languages at Kyoto University. Fujiwara has visited India several times and, as one Japanese authority on World War II remarked, "He would have no trouble getting elected in India."

Some former *Minami Kikan* and 15th Army veterans similarly retain close ties with post-war Burma under Ne Win. Kawashima Takenobu and Takahashi Hachirō, who were associated with the BIA until the end of the war, are still employed by the Burmese Government in the Burma Mission in Tokyo. Takahashi has also worked in the Burma Historical Research Insitute in Rangoon, where he translated documents on the BIA from Japanese. These men, like Col. Takeshita of the 15th Army staff, have visited post-war Burma, where they are cordially welcomed by Ne Win. Ne

Win has returned their visits.

Yanagawa of the *Beppan* still lives in Java. When the author interviewed him he was in the Indonesian Army Hospital in Djakarta. Like the three hundred other Japanese who remain in Indonesia, Yanagawa has not been able to turn his back on his wartime association with Indonesia but remains emotionally committed. The other Japanese who remain there by and large joined the Indonesian war of independence at the end of the war. Most of them were infantrymen rather than officers. They are today small businessmen scattered throughout Java and Sumatra, most of them with Indonesian wives. A few of them have revisited Japan since the war but they have no desire to return home to Japan permanently, instead regarding Indonesia as home.

Of the Japanese who trained Southeast Asians or were in the military administration bureaucracies, many have become businessmen in post-war Japan, as is true of Japanese officers generally. Most of them, if not actually involved in trade or investment in Southeast Asia, retain warm and romantic feelings about parts of Southeast Asia that they knew, and have re-visited Southeast Asia or aspire to do so. Some of these men continue wartime contacts with particular individuals they were close to.

Japanese military training programs in Asia had a total scope and impact which have heretofore escaped notice, or at least scholarly analysis. From the formation of para-military youth groups in all of Asia, to the organization and training of volunteer and independence armies, and finally to sending select officers from these indigenous armies to Japan for further training at the Military Academy, the total effect was of major significance for new and independent nations of post-war Asia. Though these armies were created to further Japanese military and political goals in wartime, the inadvertent but greater impact has been that a whole generation of Asian leadership whose nationalism was already incubating was given rigorous military training and discipline. This military experience equipped officers and men to wage wars of independence against returning Western colonial powers. It also gave them the expertise to staff and train armies of newly independent nations. Ne Win in Burma, Suharto in Indonesia, and Park Chong-hee in Korea are all products of Japanese-sponsored units and schools.

Carefully selected individuals sent to the Military Academy in Japan during the war included sixty Burmans (plus ten at the Air Force Academy), thirty-five Indians, fifty-three Filipinos, one Malay, seventeen Indonesians, and some Thais. The effectiveness of the selection process is attested by the fact that so many of these individuals today hold leading positions of military and/or political power.

Though some Japanese intelligence officers were prescient about their fostering of independence armies, most officers in Tokyo and even in Southeast Asia did not foresee that their military training would have such far-reaching results. Army officers in Southeast Asia today – though they may not always be eager to admit it – do acknowledge their debt to Japanese military training.

Japanese military occupation interjected a stimulating influence into nationalist revolutions in Southeast Asia. Educated nationalist leaders were given training to enable them to support and assist the Japanese military effort by defending themselves. Since the enemies of Japan and of Southeast Asian nationalists were identical it was possible to foster limited cooperation against Western colonial powers. In three instances at least an élite officer corps was fostered and trained. We see in Southeast Asia a dual transformation or double revolution which took a generation of young nationalists through a metamorphosis as military officers, and almost simultaneously politicized them. This is not to suggest that revolution would not have occurred had Japan not conquered and occupied Burma, Indonesia, or other parts of Southeast Asia during World War II. But certainly without the catalytic force of the Japanese occupation it would have been a somewhat attenuated and retarded revolution. Japan lent a military arm and immediacy to revolutions already in the making.

From a Japanese military perspective these revolutionary armies cannot be regarded as an immediate unqualified success. They took no independent military action apart from their Japanese mentors, at least not in the direction intended. Two of the independence armies did act on their own in revolting against their Japanese tutors. The Japanese model in this regard provided a stimulus analogous to the British impact on South and Southeast Asian nationalism, but within the telescoped period of three and a half years. The Japanese occupation provided a "positive" example in its modernizing or creative aspects, and a "negative" example in its repressive features. In both cases, however, the result was to stimulate nationalist aspirations for independence.

The Japanese advent in Southeast Asia provided the opportunity and tutelage for the creation of several armies which, while they did not by themselves shake the foundations of colonial régimes, in at least one case – the INA and the Red Fort Trial – forced a reassessment of the assumption of universal loyalty of the armed forces to the colonial power.[10] And in each area the Japanese experience and defeat created an atmosphere which inhibited a return of Western colonial rule.

It cannot be argued tenably that the Japanese Army had basically revolutionary goals for Southeast Asia in World War II, but neither can it be denied that the Japanese occupation stimulated many endemic revolutionary forces. Part of this was inadvertent, but part of it also derived from the Japanese political goal of encouraging anti-colonial independence movements.

Within the Japanese military establishment, fostering anti-colonial independence sentiment was conditional and not subscribed to by all echelons to the same degree. As early as 4 October 1940, one of the major goals as stated in the "Tentative Plan for Policy toward the Southern Regions" was to instigate independence movements in order to effect detachment of Burma, Singapore and India from British control.[11] Repeated propaganda pronouncements from Tokyo called for an end to Western imperial control in Asia, for an Asia for Asians. Japan's encouragement of Southeast Asian independence was intentional also from the viewpoint of a few individuals in intelligence agencies deputed to

[10] For a discussion of this point see Ghosh, K.K., *The Indian National Army: Second Front of the Indian Independence Movement.*

[11] International Military Tribunal for the Far East, no.638, Doct.837A.

contact nationalist leaders. Lieutenant-General Inada was a notable case within Southern Army Staff Headquarters, too, of giving impetus to the formation of volunteer forces. To these individuals should be added encouragement from scattered staff officers in local armies. Here the picture is mixed. With independence armies such as the BIA and INA the complexity of Japanese intentions is apparent. The same is true of Japan's independence policy with regard to Java and *Peta.* Japanese attitudes toward independence armies afford a revealing case study of Japanese policy toward Southeast Asian independence movements generally. An increment of romantic idealism of intelligence officers and propaganda from Tokyo is counterpoised against hard-headed military realism of operations staff officers in the Southern Army and local armies, as well as in Tokyo. IGHQ in Tokyo, while it clearly subordinated political goals to operational campaign objectives, was often willing to make propaganda value of professions of sympathy with anti-colonial movements. When apprehension was aroused by the increasingly independent and forceful behavior of nationalist leaders such as Hatta in Indonesia and Ba Maw in Burma, there were alleged *Kempeitai*-inspired plots on the lives of these leaders.[12]

The degree of Japanese support of "independence" and the nature of that "independence" should also be considered further. Burmese and Philippine independence were limited in character, qualified by Japanese policy. Tokyo did not consider deeply the ideological and political consequences of this independence, or the contradictions inherent in the notion of "qualified independence". Japanese recognition of Burmese "independence" in 1943, for example, acted as a tremendous stimulus to forces already operating in the Thakin Party since at least 1930. Anti-colonialism and nationalism were irreversible and could not be delimited in Tokyo any more than they can be today in Washington.

As earlier noted, Japanese military reverses, in 1944 and 1945 especially, produced some changes in policy in occupied areas. One of the important effects considered here

[12] Ba Maw, *Breakthrough in Burma,* pp.360-5; Kanahele, *The Japanese Occupation in Indonesia,* p.94.

was the training of guerrilla units throughout Southeast Asia and the recruitment of *Giyūgun* in several areas. These measures were taken with a view to shoring up deficient Japanese defensive strength in face of anticipated Allied landings. In Burma there were numerous last-minute proposals to increase the size of the BNA for the same reason. In many cases the Japanese also made increasing concessions to nationalist demands and in general gave less attention to maintaining the reins of control as strictly as they had early in the occupation. For Indonesia the September 1944 announcement of independence, though to specific date was mentioned, came as a qualified concession.

Japan's total collapse of course had a critical importance for independence movements everywhere in Southeast Asia. It meant that real independence was an imminent possibility, that the return of Western colonial rule was also a possibility not to be tolerated. All the self-confidence, military training, and political expertise which had been acquired in the Japanese interregnum were now mobilized against attempts at reassertion of Western colonial rule. Under Japanese occupation nationalism and aspirations for independence had been stimulated to a point of no return. This was a lesson the British and Dutch had to learn in the post-war years.

The myth of colonial omnipotence was exposed both by the defeat of Western colonial powers by the outnumbered Japanese and then by Japan's own collapse. The Japanese interregnum marked the end of any claim by Western colonial authority to rule in Southeast Asia. Japan's encouragement of independence movements, arming of Southeast Asian nationalists, and then her defeat left local nationalists with the confidence that they would never again be compelled to submit to foreign rule, and the means to ensure that it did not happen. Southeast Asians emerged from the war equipped with military arms and training, political and organizational experience and skills, and a firm and irrevocable ideological commitment to independence. Had Japan succeeded militarily, the impact of Japanese occupation on independence movements would have been quite another story.

These gains were not without a price. The end of Western

colonial rule was effected in a manner that left confusion and disruption of European economic activity in the immediate post-war years. In some cases recovery was a long and arduous process. In Burma, for example, not until 1957 were pre-war production levels attained. In Indonesia pre-war production levels were reached in 1953, but within a year civil war between Java and the Outer Islands again created chaos.

Political divisions among nationalists in both Indonesia and Indochina made it possible for the Dutch and French to return temporarily after the war and for the Dutch to carry out "police actions". These attempts at colonial reassertion evoked a bitter response among the Viet Minh and the Indonesians. Attempts at mediation by other powers brought both conflicts to prominence within the international arena.

Beyond the palpable military value of Japanese training for Southeast Asians has been the political aftermath of the Japanese experience. Military training coupled with explosive nationalism provided a potent new resource for the leadership of post-war Southeast Asia. In all the Japanese-occupied nations where independence and volunteer armies were trained, an officer corps was also politicized during the war, whether as a direct or indirect result of the Japanese intent. The INA officers were exceptional in rejecting their pre-war military career profiles and turning almost completely to politics. All other Japanese-trained armies were, in the process of acquiring military discipline and experience, also politicized and imbued with a new sense of self-confidence and national self-esteem. This new sense of national political identity has conditioned the emergence of the military-political leadership of post-war Southeast Asia. Japan's wartime impact thus continues in contemporary Southeast Asia.

Appendices

(A) Statement to Colonel Suzuki on the eve of his departure from Burma

Burma Independence Army
Bestowal of Blessing (and) Valedictory

We Burmese people through 2500 years of history have established successive kingdoms and their great capital cities such as Tagaung, Tharay-Kitiya, Myin-Saing, Pinya, Ava, Sagaing, Amarapura, Gon-Baung, Toungoo, and Mandalay. With Our Own Royal Throne and the Royal Insignia, The Golden Umbrella, we have lived in golden splendour and dignity. We are descended from the earliest monarchs, the Ancient Kings known as Maha-Thammada Min, from whom sprang the unbroken succession of appointed kings of the Sacred Castle of Thaki-Wun; they have ruled over our independent and sovereign nation whose civilization has made contribution to the history of the world. We belong to a master race.

Even so, however, Burma suffered a fateful weakening of her horoscope and in that time of weakness, over 100 years ago, Westerners called English who came from 8000 miles away unlawfully seized Burma and consigned the Burmese people into slavery. Using myriad unholy schemes, the English repressed and restricted the Burmese people who suffered the destruction of their culture, health, education, economy and religion.

The infidel [note: word used here is literally translated into "against the Dharma" – Dharma means the Buddhist religion. Alternative word to "infidel" is "unholy"] English did not confine themselves to our Burma. They monopolized all the countries of Asia. The national treasures, rich mineral resources and bountiful agricultural produce of Asian lands were greedily plundered by this unlawful usurper. It was as though the people of Asia were plunged into the Four Kinds of Darkness [note: these 4 are: 1) darkness of night without moonlight; 2) darkness of dense forests; 3) darkness due to thick clouds; and 4) darkness at time of midnight] and they were without support from any quarter. Forced thus into absolute darkness, the people of Asia did not know how long it might be before the light of law and justice would dawn again.

During that time of great darkness for the helpless people of Asia, the Japanese of Asia alone raised the morale and spirit of the people of Asia by their previous victory 37 years ago over the mightily armed Russians when, for the first time, the rays of liberty shone upon the Asian continent.

The Japanese, the vanguard of the Asian people, struggled for social, educational and economic progress not only for their own people but also for the politically and economically oppressed people in chains in Asian countries like India, Burma, China, Malaya, the Philippines, Sumatra, etc., by swiftly driving the devil English and Americans from these Asian lands. They liberated these various Asian countries and established a new and modern era. Included in their great and unceasing supreme endeavour was the liberation of Burma. To this end, Mi-Nan-Mi known as General Mo Gyo [note: Mo Gyo in Burmese means "Thunderbolt" – Gyo is pronounced in the same way as Jo], with the help of Bo Dé Za known as Thakin Aung San who headed the Band of Heroic Young Men, organized and established the Burma Independence Army which marched into Burma with the Imperial Army. The enemy, English, American and Chinese soldiers were rapidly defeated and annihilated. Within 70 days the remnants of these devils were driven completely out of the country; their retreat was indeed so precipitous that they had no time even to put their trousers on. The entire world is aware of this victory and it will be hailed in the historical annals of the Burmese Independence Movement.

Just as a father instructs his beloved sons and daughters to refrain from evil and to do good, and just as he protects them and cares for them, General Mo Gyo has instructed, protected and cared for all members of the Burma Independence Army with true and sincere affection, and the men and women of Burma will never forget this.

To become a secure and independent nation in the world today, Burma cannot depend solely on a land army like the Burma Independence Army. Therefore, General Mo Gyo began organizing the Burma Navy.

General Mo Gyo who has labored mightily for Burma is about to return to Japan. We will always cherish the father figure, patron and benefactor of the Burma Independence Army. Also, his endeavours on behalf of the Burmese nation will be gratefully remembered. Even though the world itself may disappear, our feelings of gratitude towards him will never disappear.

When the General returns to Japan, please report and convey what you have personally experienced of our hospitality, loyalty, courage, our assistance to Japanese soldiers, and our efforts towards Japanese-Burmese friendship and close co-operation, to the Japanese Emperor, Prime Minister Tōjō, top level leaders and the men and women of Japan. In order that Burma may be able to gain independence rapidly:

a) The Burma Independence Army created by General Mo Gyo should be strengthened and expanded with more arms, equipment and personnel.
b) The Burma Navy requires ships, large and small, to increase its effectiveness.
c) We sincerely belief that before long the Burma Air Force will be created.

May General Mo Gyo from this day on be blessed with continuing wealth and prosperity. May he be free from harm wherever he may travel. May he be blessed with peace and happiness and with long life [note: literal translation of Burmese phrase wishing long life is: "May you/he live to be over 100 years"]. We of the Burma Independence Army pray for you and will always treasure your memory. In order that the General in return keep us, the Burma Independence Army, in his memory, we most respectfully and humbly present to him this valedictory encased in this ornate and magnificant casket of silver.

Burma Independence Army
Burma

Translated by Sao Ying Sita Naw Hseng Lao

ဗမာပြည်

ဗမာ့လွတ်လပ်ရေးတပ်မတော်

ထုတ်မင်္ဂလာ

-ပြဿနာ-

ကျွန်ုပ်တို့ဗမာလူမျိုးများသည်လွန်ခဲ့သည့်နှစ်ပေါင်း - ၂၅၀၀-ကျော်က စချ၍-တကောင်းသရေခေတ္တရာ-မြင်ဆိုင်းပင်းယ-အင်းဝ-စစ်ကိုင်း-အမရပူရ-ကုန်းဘောင်-တောင်ငူ-မန္တလေး စသောမြို့ကြီးများတွင်ရွှေထီးရွှေနန်း- ရွှေကြာငှန်း စိုက်ထူ၍-မဟာသမတမင်းမှဆင်းသက်၍အသဘိန္တခတ္တိယ နွယ်စင်သာကီဝင်မင်းအဆက်ဆက်တို့ဖြင့်မပျက်နွယ်ရိုးစိုးမိုးအုပ်ချုပ်၍လွတ်လပ်သောလူမျိုးတမျိုးအဖြစ်ဖြင့်ကမ္ဘာပေါ်တွင်ရာဇဝင်ပြောင်ခဲ့သောယဉ်ကျေးသည့်နိုင်ငံကြီးသားသခင်လူမျိုးများဖြစ်ခဲ့ပေသည်။

သို့ပင်ဖြစ်လင့်ကစားဗမာပြည်၏ကံဇာတာသည်-ဖန်လာညှိုးမှိန်အချိန်အခါမသင့်သဖြင့်-မိုင်ပေါင်း ၈၀၀၀-ကျော်ကွာဝေးသောအနောက်နိုင်ငံသားအင်္ဂလိပ်များသည်လွန်ခဲ့သည့်နှစ်ပေါင်း ၁၀၀-ကျော်က ကျွန်ုပ်တို့၏ ဗမာပြည်ကိုမတရားတိုက်ခိုက်သိမ်းပိုက်ပြီးသော် - ကျွန်ုပ်တို့ဗမာလူမျိုးများအား ကျွန်စရင်းသို့သွတ်သွင်း၍အမျိုးမျိုးသောအစီအမံနည်းတို့ဖြင့်ဗမာလူမျိုးတို့၏ယဉ်ကျေးမှုကိုငှင်းတုံ့နှာရေး-ပညာရေး-စီးပွားရေး-သာသနာရေး စသည်တို့ကိုငှင်း ဖိနှိပ်ချုပ်ချယ်၍ခေါင်းမထောင်နိုင်အောင်အဘက်ဘက်မှဖျက်ဆီးချိုးနှိမ်ခဲ့လေသည်။

အစမှသမားအင်္ဂလိပ်များသည်ကျွန်ုပ်တို့၏ဗမာနိုင်ငံသာမကအာရှတိုက်တတိုက်လုံးရှိတိုင်းနိုင်ငံအနှံ့အပြားတွင်လက်ဝါးကြီးအုပ်၍အရှေ့နိုင်ငံများ၏အဘိုးတန်ရတနာ-ဓါတ်သတ္တု-မြေဆီဩဇာအဆီအနှစ်စသည်တို့ကို။မြိန်ရေရှက်ရေစုတ်ယူ၍မတရားမင်းမူလျက်ရှိသောကြောင့်အာရှတိုက်တတိုက်လုံးရှိအရှေ့နိုင်ငံသူနိုင်ငံသားအပေါင်းတို့မှာ။အမိုက်တိုက်လေး ပါးသို့ဝင်သည်နှင့်မခြား။အားကိုးရာမရှိ။ပကတိအမိုက်မှောင်အတွင်းသို့သက်ဆင်းကျရောက်လျက်ရှိခဲ့ပေရာ။မည်သည့်အချိန်အခါရောက်ခါမှ။တရားမှန်လမ်းမှန်တည်းဟူသောအလင်းရောင်ကိုရပါအံ့နည်းဟူ၍ခန့်မှန်းမထင်၊မသိမမြင်နိုင်အောင်ရှိခဲ့ရပေသည်။

ထိုသို့အာရှတိုက်သားအပေါင်းတို့မှာမျက်ကန်းအသွင်။အားကိုးရာမမြင်ဖြစ်နေစဉ်အတွင်း။အာရှတိုက်သားဂျပန်သည်။ဗိုလ်ပါအင်အားများပြားလှသောရုရှားလူမျိုးတို့အားလွန်ခဲ့သည့်နှစ်ပေါင်း ၃၇ နှစ်ကတိုက်ခိုက်အောင်မြင်ခဲ့ရာတွင်မှ။အာရှတိုက်သားတို့သည်-စိတ်အားတက်ကြွကာလွတ်လပ်ရေးရောင်ခြည်သည်အာရှတိုက်တခွင်တွင်ကွန့်မြူးရက်သန်းလာသည်ကိုတွေ့မြင်ရပေကြောင်း။

ထို့နောက်အာရှတိုက်သားတို့၏ရှေ့ဆောင်သဘွယ်ဖြစ်သောဂျပန်အမျိုးသားတို့သည်မိမိတို့၏လူမျိုးရေး။စီးပွားရေး၊ပညာရေးစသောတိုးတက်ကြီး

ပွားများတို့ကိုသာအဓိကထား၍ ကြိုးပမ်းဆောင်ရွက်သည်မဟုတ်ဘဲ။ နိုင်ငံရေးအချုပ်အချယ်။
စီးပွားရေးအချုပ်အချယ်တို့ဖြင့်။ မလူးသာမလွန့်သာအောင်။ တုပ်နှောင်ဖွဲ့စည်းထားခြင်းခံ
ရသော။ အိန္ဒိယပြည်။ ဗမာပြည်။ တရုတ်ပြည်။ ပရှူးကျွန်းစွယ်။ ဖိလစ်ပိုင်ကျွန်း။ စုမတြာ
ကျွန်းအစရှိသော အရှေ့ဘက် နိုင်ငံအသီးသီးတို့မှ မိစ္ဆာကောင်များဖြစ်သော အင်္ဂလိပ် နှင့်
အမေရိကန်တို့ကို-ဒလကြမ်း မောင်းနှင်ထုတ်ကာ အာရှတိုက်ရှိ နိုင်ငံအသီးသီးအား လွတ်
လပ်ရေးရစေလျက်ခေတ်သစ်ထူထောင်ပေးရန်။ လေးလေးစားစား။ မနေမနား အားထုတ်ကြိုး
ပမ်းရာတွင်။ ကျွန်ုပ်တို့ဗမာပြည်၏လွတ်လပ်ရေးနှင့်ခေတ်သစ်ထူထောင်ရေးတို့လည်း
စွမ်းစွမ်းတမန် ဆောင်ရွက်နိုင်ရန် မိနံမိအမည်ရှိသော ဗိုလ်ချုပ်ကြီး-ဗိုလ်မိုးကြိုးသည်-
ဗိုလ်တေဇ အမည်ခံသခင်အောင်ဆန်း အမှူးရှိသော အာဇာနည် ယောက်ျားလူငယ်များ၏
အကူအညီဖြင့်။ ဗမာ့လွတ်လပ်ရေးတပ်မတော်ကြီးကို။ ကြီးမှူး ဦးဆောင်ကာ တည်
ထောင်ဖွဲ့စည်းလျက်။ ဂျပန်ဘုရင့်တပ်မတော်များနှင့်အတူ။ ဗမာပြည်အတွင်းသို့ချင်းနင်း
ဝင်ရောက်ခဲ့ပြီးသော်ရန်သူဖြစ်သော - အင်္ဂလိပ် - အမေရိကန် နှင့် တရုတ် စစ်သည်များ
အားထုံးလိုချေရေလိုနှောက် ခါ။ တိုက်ခိုက် နှိမ်နင်းလျက်-ကြွင်းကျန်သော မိစ္ဆာကောင်များ
အားမြန်မာပြည်မှ အပသို့ရောက်ပေါင်း [illegible] အတွင်း [illegible] မျှအစိုး
မရစေဘဲ ဒလကြမ်းမောင်းနှင်ထုတ် ခဲ့သည်ကိုတကမ္ဘာလုံး သိရှိပြီးဖြစ်သည့်အတိုင်းဗမာ့
လွတ်လပ်ရေးရာဇဝင်တွင်မော်ကွန်းတင်ရမည်ဖြစ်သော အကြောင်းကြီးတရပ်ဖြစ်ပေသည်။

အဖဖြစ်သူသည်မိမိ၏ ရင်နှစ်။ သားချစ်သမီးတို့အား မကောင်းမြစ်တား-ကောင်း
ရာညွှန်ပြဆုံးမသွန်သင်ကာ စောင့်ထိမ်းအုပ်ချုပ်ဘိ သကဲ့သို့- ဗိုလ်ချုပ်ကြီး ဗိုလ်မိုးကြိုး
သည်လည်း၊ ဗမာ့လွတ်လပ်ရေးတပ်မတော်သားများအား ထူးကဲမွန်မြတ်သော စေတနာ
ထားကာ ဆုံးမသွန်သင်ပဲ့ပြင်လျက် အုပ်ထိမ်းစောင့်ရှောက် ခဲ့သည်ကိုလည်း ဗမာပြည်
သူပြည်သားအပေါင်းတို့သည်မည်သည့်အခါမျှမေ့လျော့နိုင်လိမ့်မည်မဟုတ်ပေ။

ထိုမှတပါး ကမ္ဘာပေါ်တွင်လွတ်လပ်သော နိုင်ငံတခုအဖြစ်ဖြင့် တည်တံ့ခိုင်မြဲ
နိုင်ရန်အရေးမှာ။ ဗမာ့လွတ်လပ်ရေးတပ်မတော်ကြီးကဲ့သို့ ကုန်းတပ်ကြီးတခုတည်း
ဖြင့်နိုင်လုံစိတ်ချရသည်မဟုတ်သောကြောင့်။ ဗမာ့လွတ်လပ်ရေး - ရေတပ်မတော်ကြီး
ကိုလည်းမကြာမီကပင်-ဗိုလ်ချုပ်ကြီး ဗိုလ်မိုးကြိုးသည် စတင်ဖွဲ့စည်းပေးခဲ့ပေသည်။

ကျွန်ုပ်တို့သျှင်ဗမာပြည်၏အကျိုးကို အားကြိုးမာန်တက် - ဆောင်ရွက် သည်ဖိုး
ခဲ့ပေသော ဗိုလ်ချုပ်ကြီးဗိုလ်မိုးကြိုး သည်မကြာမီ ဂျပန်ပြည်သို့ပြန်လည်တော့
မည်ဖြစ်ရာ ဗမာ့လွတ်လပ်ရေး တပ်မတော်ကြီး၏ မိဘအသွင်၊ ကျေးဇူးရှင်ဗိုလ်ချုပ်
ကြီးအားကျွန်ုပ်တို့သည်အမြဲအစဉ်သတိရမည်ဖြစ်သည့်ပြင်- ဗမာပြည်အတွက်ဗိုလ်
ချုပ်ကြီးဆောင်ရွက်ခဲ့သမျှသော ကျေးဇူးတရားတို့ သည် ကမ္ဘာကြေသော်လည်းဥဒါန်း
ကြေတော့မည်မဟုတ်ပေ။

ဂျပန်ပြည်သို့ ဗိုလ်ချုပ်ကြီး ပြန်လည်ဆိုက်ရောက်သောအခါ၌လည်းကျွန်ုပ်
တို့ ဗမာလူမျိုးအပေါင်း၏ ညည့်ဝတ်ကျေပုံ။ သစ္စာရှိပုံ။ သူရသတ္တိ နှင့်ပြည့်စုံပုံ။
ဂျပန်စစ်သည်တော်များအားမည်သို့ကူညီပုံ။ ဂျပန်နှင့် ဗမာတို့ ချစ်ခင်ရင်းနှီးစွာ
ပေါင်းသင်းဆက်ဆံပုံတို့ကိုဗိုလ်ချုပ်ကြီး မြင်တွေ့ရသည့်အတိုင်း။ ဂျပန်ဘုရင်မင်း
မြတ်နှင့် တကွနန်းရင်းဝန်တိုဂျိုစသော ထိပ်သီးခေါင်းဆောင်များနှင့် ဂျပန်အမျိုး

သား-အမျိုး သ္မီးတို့အားတဆင့်စကားပြောကြားကာ။ ဗမာပြည်၏လွတ်လပ်ရေး
ကိုမနှေးအမြန်ရရှိစေရန်အတွက်

(က) ဗိုလ်ချုပ်ကြီး ဦးစီး တည်ထောင်ခဲ့သော ဗမာ့လွတ်လပ်ရေး
တပ်မတော်ကြီးလည်း ယခုထက်-လက်နက်အင်အားပြည့်စုံတောင့်
တင်းစေရန်။

(ခ) ဗမာ့လွတ်လပ်ရေး-ရေတပ်မတော်ကြီးလည်း စစ်သင်္ဘောကြီး
ငယ်တို့ဖြင့်တင့်တယ်ခန့်ငြားစွာ-များများမတ်မတ်ရပ်တည်နိုင်
စေရန်။

(ဂ) မကြာမီလည်း-ဗမာ့လွတ်လပ်ရေး-လေတပ်မတော်ကြီးတည်
ထောင်ဖွဲ့စည်းပေးစေရန်-ကြံဆောင်ကြိုးစားလိမ့်မည်ဟု ကျွန်ုပ်တို့
သည်အထူးမျှော်လင့်ယုံကြည်လျက်ရှိနေပါကြောင်း။

ဗိုလ်ချုပ်ကြီး ဗိုလ်မိုးကြိုးသည်ယနေ့မှ စ၍ ဥစ္စာ စည်းစိမ် ဂုဏ်သိမ်ဒြပ်
တို့သည် တကံသစ် စနေဝန်း။ ထွန်းသစ်စလကဲ့သို့တနေ့တခြားတိုးတက်ကြီးမြင့်
ကါ။ သွားလေရာ။ လာလေရာအရပ်တို့၌ ဘေးမထိရန်မသန်းဘဲ ချမ်းသာယာပြီးသော်။
သက်တော်ရာကျော်ရှည်ပါစေသော်ဟူ၍။ ကျွန်ုပ်တို့ဗမာ့လွတ်လပ်ရေးတပ်မတော်
ကြီးကလှိုက် လှိုက် လှဲ လှဲ ဆုတောင်းပတ္ထနာပြုပြီးသော်။ ကျွန်ုပ်တို့သည်ဗိုလ်ချုပ်ကြီး
အားအမြဲအစဉ်သတိရဘိ သကဲ့သို့။ ဗိုလ်ချုပ်ကြီးကလည်း-ကျွန်ုပ်တို့ဗမာ့လွတ်လပ်
ရေးတပ်မတော်ကြီးအားအမြဲအစဉ်သတိရစေရန်။ ဤထုထိမင်္ဂလာပြဿဘာစါတမ်း
ကိုကြက်သရေရှိလှသော-တင့်တယ်လှပသည့်ငွေကျုပ်တွင်-သွတ်သွင်း၍သက္ကစ္စ
ဂါရဝဖြင့်ရိုသေစွာဆက်ကပ်ခြင်းပြုပါသတည်း။

ဗမာ့လွတ်လပ်ရေးတပ်မတော်
ဗမာပြည်။

(B)

Estimated Size of Japanese-trained Armies

Indian National Army	35-40,000
Burma Independence Army	200,000 (recruits, not trained)
Burma Defense Army	4,000 expanded to 55,000
Peta Java	33,000
Bali	1,500
Malaya *Giyūgun*	2,000
Malaya *Giyūtai*	5,000
Sumatra *Giyūgun*	5-6,000
Philippines *Makapili*	6,000
Indochina *Giyūgun*	3,000
Borneo *Giyūgun*	1,3-1,500

Southeast Asians trained at the Shikan Gakkō (Military Academy)

Indians	35
Burmans	60, plus 10 at the Air Force Academy
Thais	several
Malays	1
Indonesians	17
Filipinos	53

Bibliography

Japanese Sources

Bōeichō Bōei Kenshūjo Senshishitsu (Defense Agency, Defense Training Institute, War History Library), *Biruma kōryaku sakusen* (Burma offensive operation), Tokyo, 1967.

Bōeichō Bōei Kenshūjo Senshishitsu, *Imparu sakusen* (The Imphal operation), Tokyo, 1968.

Bōeichō Bōei Kenshūjo Senshishitsu, *Irawaji kaisen – Biruma bōei no hatan* (The Irawaddy battle – failure of the defense of Burma), Tokyo, 1969.

Bōeichō Bōei Kenshūjo Senshishitsu, *Marē shinkō sakusen* (Malaya offensive operation), *Tokyo, 1966.*

Bōeichō Bōei Kenshūjo Senshishitsu, *Shittan Mei-go sakusen* (The Sittang, "Operation Mei"), Tokyo, 1970.

Bōeichō Bōei Kenshūjo Senshishitsu, *Ran-In kōryaku sakusen* (Netherlands Indies offensive operation), Tokyo, 1967.

Dai Hon'ei seifu renraku kaigi kettei gijiroku (Records of Imperial General Headquarters-Government Liaison Conference Decisions), 1941-45.

Dai Jūgogen Shireibu (15th Army Headquarters), *Biruma ni okeru gunsei meirei setsu* (Compilation of military administration orders in Burma), June-October 1942, Maimyo.

Fujiwara Iwaichi, *F Kikan* (F Agency), Tokyo, 1966.

Fujiwara Iwaichi, *F Kikanchō no shuki* (Memo of the 'F' Agency chief), Tokyo, 1959.

Gaimushō (Foreign ministry), *Dai Tōa sensō kankei ikken; Indō mondai* (Matters relating to the Greater East Asia War; the India problem), 1943-44.

Gaimushō (Foreign ministry), *Dai Tōa sensō kankei ikken; Indō mondai* (Matters relating to the Greater East Asia War; the India problem), 1943-44.

Gaimushō, *Dai Tōa sensō kankei ikkan; Biruma mondai* (Matters relating to the Greater East Asia War; the Burma problem), 1941-45.

Gaimushō, *Nihon gaikō nempyō narabi ni shuyō monjo* (Chronology of Japanese diplomacy together with important documents), 2 vols., Tokyo, 1966.

Gaimushō, Ajiya Kyoku (Asia Office), *Subasu Chandora Bosu to Nihon* Tokyo, 1956.

Gendai shi shiryō (Modern historical materials), Misuzu Shobo, ed., 43 vols., Tokyo, 1962-70, especially vol. 1, *Nitchū sensō* (The Sino-Japanese War).

Hattori Takushirō, *Dai Tōa sensō zenshi* (Complete history of the Greater East Asia War), Tokyo, 1965 (one vol. ed.).

Iida Shōjirō, *Senjin yūwu* (Twilight battlefield tales), privately published, Tokyo, 1967.

Imamura Hitoshi Taishō kaisōroku (Memoirs of General Hitoshi Imamura), 10 vols., mimeo., also published in 4 vols. by Jiyū Ajiya Sha, Tokyo, 1960, especially vol. 4, *Tatakai o owaru* (The Struggle Ends).

Ikeda Yū, ed., *Hiroku Dai Tōa senshi* (Secret history of the Great East Asia War), 6 vols., Tokyo, 1954, especially *Mare-Biruma hen* (Malaya-Burma volume).

Imai Takeo, "Ajiya dokuritsu ni hatashita meishu Nihongun no kōzai" (The merits and demerits of Japanese Army leadership in achieving Asian independence), *Maru*, v.XX, no.9, Sept. 1967, pp.212-19.

Imaoka Yutaka, *Nanseihōmen rikugun sakusenshi* (Campaign history of the Southwest area army), 1944, mimeo.

Inada Masazumi, *Shōnan nikki* (Singapore diary), 3 vols. summary, mimeo.

Ishii Akiho, *Nampō gunsei nikki* (Diary of military administration in the Southern Theater), Nov. 1941-Jan. 1943, mimeo.

Isoda Saburō, *Isoda Saburō Chūjo kaisōroku* (Recollections of Lieutenant-General Isoda Saburō), 1954, mimeo.

Itagaki Yoichi, *Ajiya no taiwa* (Conversations with Asia), Tokyo, 1968.

Izumiya Tatsurō, *Biruma dokuritsu hishi; sono na wa Minami Kikan* (A secret history of Burmese independence; its name, the *Minami Kikan*), Tokyo, 1969.

Kaneko Seigo, "Annam himitsu butai" (Annam secret unit), *Shūkan Yomirui, Nihon no himitsu sen (Weekly Yomiuri,*

Japan's secret war), special issue, December 8, 1956.

Kawabe Shōzō, *Biruma nikki shōroku* (Excerpts from the Burma diary), 1944, mimeo.

Kokubu Shōzō, *Biruma seiji undō no tembō* (A survey of Burmese political movements), mimeo., Nanyō Keizai Kenkyūjo (South Seas Economic Research Institute), 1942.

Kokubu Shōzō, *Dai Biruma shi* (History of Greater Burma), 2 vols., Tokyo, 1944.

Kokubu Shōzō, "Independence Draft Plan" in Gaimushō, *Dai Tōa sensō kankei ikken; Biruma mondai,* (Matters concerning the Greater East Asia War; the Burma problem) *s.d.*

Kunizuka Kazunori, *Indōyō ni kakeru niji* (Rainbow over the Indian Ocean), Tokyo, 1958.

Kurasawa Aiko, "*Peta* to 45 nen no kakumei" (*Peta* and the 1945 revolution) M.A. thesis, Tokyo University, 1971.

Kurasawa Aiko, "Historical considerations concerning the Blitar Revolt – toward the origins of Indonesian independence" (in Japanese), graduation thesis, Tokyo University, 1969.

Kuroda Hidetoshi, *Gunsei* (Military administration), Tokyo, 1952.

Machida Keiji, *Tatakau bunka butai* (A fighting propaganda unit), Tokyo, 1967.

Maruyama Shizuo, *Nakano Gakkō – tokumu kikan-in no shuki* (The Nakano School – memoir of a member of a special agency), Tokyo, 1948.

Masuda Atō, *Indonejiya gendaishi* (Modern Indonesian history), Tokyo, 1971.

Min Gaung, Bo, "Aung San shogun to sanjunin shishi", (General Aung San and the thirty comrades), *Shiroku* (Historical record), no. 3, Kagoshima, 1970.

Miyamoto Shizuo, *Jawa shūsen shoriki* (Records of the settlement of the end of the war in Java), Tokyo, 1973.

Miyoshi Shinkichirō, "Jawa senryō gunsei kaikoroku" (Recollections of military administration in occupied Java), *Kokusai mondai* (International problems), nos.61-82, 15 parts, 1956-66, Tokyo.

Murata Heiji, *Imparu sakusen – Retsu heidan Kohima no shitō* (The Imphal operation – Kohima death struggle of the Retsu group), Tokyo, 1967.

Naka Eitarō, comp., *Biruma gunsei shi* (History of Burmese military administration), 4 vols., 1943.

Nakamiya Gorō, "Sumatra ni okeru muketsu senryō no kage ni" (Behind the bloodless occupation of Sumatra), *Shukan Yomiuri, Nihon no himitsu sen (Weekly Yomiuri,* Japan's secret war), 1956.

Nampō sakusen ni tomonau senryōchi gyosei no gaiyō (Outline of administration in occupied areas pursuant to the Southern offensive), mimeo., 1946.

Ohno Tohru, "Biruma kokugunshi" (History of the Burmese national army), *Tōnan Ajiya kenkyū* (Southeast Asian Studies), vol. 8, 1970, pts.1-3.

Ōta Tsunezō, *Biruma ni okeru Nihon gunseishi no kenkyū* (The study of Japanese military administration in Burma), Tokyo, 1967.

Sambō Hombu Daiichibu Kenkyūhan (General Staff Headquarters, First Division, Research Unit), *Nampō sakusen ni okeru senryōchi tochi yōkōan* (Draft outline for control of occupied areas in the Southern offensive), 1941.

Satō Kenryō, *Dai Tōa sensō kaikoroku* (Recollections of the Greater East Asia War), Tokyo, 1966.

Sawamoto Rikichirō, *Nihon de mita Birumagun no seiritsu* (The establishment of the Burmese Army as seen from Japan), 3 vols., mimeo.

Sōma Kurohiko and Sōma Yasuo, *Ajiya no mezame* (The awakening of Asia), Tokyo, 1955.

Sugii Mitsuru, *Minami Kikan gaishi* (Unofficial history of the *Minami* Kikan), 1948, mimeo.

Takagi Sōkichi, *Taiheiyō sensō to rikukaigun no kōsō* (The Pacific War and Army-Navy rivalry), Tokyo, 1967.

Tanaka Masaaki, *Fusetsu yonjūgonen no yume: hikari mata kaeru* (A dream of vicissitudes of forty-five years: the light returns again), Tokyo, 1959.

Taya, Bo, "Sanjūnin shishi no kikoku" (The thirty comrades' return to Burma), *Shiroku,* no.3, 1970, Kagoshima.

Tanemura Suketaka, *Dai hon'ei kimitsu nisshi* (Secret journal of Imperial General Headquarters), Tokyo, 1952.

Tokugawa shiryō (Tokugawa materials), especially no.28, "Reference materials on nationality policy", marked

Sōmubu sōmuka (General Affairs Division, General Affairs Section), *n.d.*

Tsuji Masanobu, *Jūgo taiichi – Biruma no shitō* (Fifteen versus one – the death struggle of Burma), Tokyo, 1950.

Tun Hla, Bo, "Aung San shogun, Biruma dokuritsu no tateyakusha" (General Aung San, leading spirit of Burmese independence), *Kadai shigaku* (Kagoshima University, Historical studies), no.15, 1967.

Tsuchiya Kisou, *Indonejiya giyūgun to Suharto daitōryō* (The Indonesia Volunteer Army and President Suharto), Tokyo, *s.d.*, 24 pp.

Umemoto Sutezō, *Biruma hōmengun* (Burma Area Army), Tokyo, 1969.

Ushiro Masaru, *Biruma sensenki* (Burma battle record), Tokyo, 1953.

Waseda Daigaku Shakai Kagaku Kenkyūjo (Kishi Kōichi, and Nishijima Shigetada, eds.), *Indonejiya ni okeru Nihon gunsei no kenkyū* (Studies in Japanese military administration in Indonesia), Tokyo, 1959.

Yanagawa Munenari, "Jawa no Beppan" (Beppan of Java), *Shūkan Yomiuri, Nihon no himitsu sen, (Weekly Yomiuri,* Japan's secret war) Tokyo, 1956, pp.149-55.

Yanagawa Munenari, *Rikugun chōhōin Yanagawa Munenari chūi* (Lieutenant Yanagawa Munenari, an Army intelligence officer), Tokyo, 1967.

Yomiuri Shimbunsha, *Shōwashi no Tennō* (The Emperor in Showa history), 11 vols., Tokyo, 1969-71.

English Sources

Akashi, Yoji, "Japanese Military Administration in Malaya – its Formation and evolution in reference to Sultans, the Islamic Religion, and the Muslim Malays – 1941-1945", *Asian Studies* 7:(1969) 81-110, Quezon City.

Akashi, Yoji, "The Japanization Program in Malaya with Particular Reference to the Malaya", 1971, mimeo.

Akashi, Yoji, *The Nanyang Chinese National Salvation Movement, 1937-1941,* East Asian Series Research Publication, No. 5, Center for East Asian Studies, University of Kansas, Lawrence, 1970.

Allen, Louis, "Studies in the Japanese Occupation of South-East Asia, 1942-45 (I) – Ba Maw and the Independence of Burma", *Durham University Journal,* LXIII, new series XXXII:(1970) 1-15.

Allen, Louis, "Studies in the Japanese Occupation of South-East Asia, 1942-1945 (II): 'French Indo-China' to 'Vietnam', Japan, France, and Great Britain, Summer 1945", *Durham University Journal.*

Anderson, Benedict R.O.G., *Some Aspects of Indonesian Politics under the Japanese Occupation: 1944-1945,* Ithaca, 1961.

Anderson, Benedict R.O.G., *Java in a Time of Revolution, Occupation and Resistance, 1944-1946,* Ithaca, 1971.

Aziz, M.A., *Japan's Colonialism and Indonesia,* The Hague, 1955.

Ba Maw, *Breakthrough in Burma, Memoirs of a Revolution, 1945-1946,* New Haven, 1968.

Ba Than, Dhammika U., *The Roots of the Revolution, a brief history of the Defence Services of the Union of Burma and the Ideals for which they stand,* Rangoon, 1962.

Ba U, U, *My Burma, the autobiography of a President,* New York, 1959.

Benda, Harry J., "The Beginnings of the Japanese Occupation of Java", *Journal of Asian Studies,* 15: (1955-56) 541-60.

Benda, Harry J., *The Crescent and the Rising Sun,* The Hague, 1958.

Benda, Harry J., James K. Irikura, and Kōichi Kishi, *Japanese Military Administration in Indonesia: Selected Documents,* New Haven, 1965.

Berreman, Joel V., "The Japanization of Far Eastern Occupied Areas", *Pacific Affairs,* v.7, (1944) pp.168-80.

Boyle, John Hunter, *China and Japan at War, 1937-1945, The Politics of Collaboration,* Stanford, 1972.

Brett, Cecil Carter, "Japanese Rule in Malaya, 1942-1945", M.A. thesis, University of Washington, 1950.

Burchett, W.G., *Trek Back from Burma,* Allahabad, 1947.

Butwell, Richard, *U Nu of Burma,* Stanford, 1963.

Cady, John F., *A History of Modern Burma,* Ithaca, 1958.

Christian, John LeRoy, *Burma and the Japanese Invader,* Bombay, 1945.

Cohen, Stephen P., *The Indian Army, its Contribution to the Development of a Nation,* Berkeley, 1971.

Collis, Maurice Stewart, *Last and First in Burma,* London, 1956.

Dahm, Bernhard, *Sukarno and the Struggle for Indonesian Independence,* Ithaca, 1969.

Elsbree, Willard H., *Japan's Role in Southeast Asian Nationalist Movements,* Cambridge, Mass., 1953.

Evans, Sir Geoffrey, *Imphal, a Flower on Lofty Heights,* London, 1962.

Evans, Sir Geoffrey, *Slim as Military Commander,* Princeton, 1969.

Evans, Humphrey, *Thimayya of India, a Soldier's Life,* New York, 1960.

Feith, Herbert, *The Decline of Constitutional Democracy in Indonesia,* Ithaca, 1962.

Fischer, Louis, *The Story of Indonesia,* New York, 1959.

Gutteridge, William, *Armed Forces in New States,* Oxford, 1962.

Gutteridge, William, *Military Institutions and Power in The New States,* New York, 1965.

Ghosh, K.K., *The Indian National Army, Second Front of the Independence Movement,* Meerut, 1969.

Gordon, Ian Fellows, *Amiable Assassins, The Story of the Kachin Guerrillas of North Burma,* London, 1957.

Gregory, Ann, "Dimensions of Factionalism in the Indonesian Military", 1970, mimeo.

Guyot, Dorothy, *The Political Impact of the Japanese Occupation of Burma,* New Haven, Yale Ph.D. dissertation, 1966.

Hammer, Ellen J., *The Struggle for Indochina,* Stanford, 1954.

Hanna, Willard A., *Eight Nation Makers, Southeast Asia's Charismatic Statesmen,* New York, 1965.

Hartendorp, A.V.H., *The Japanese Occupation of the Philippines,* 2 vols., Manila, 1967.

Hartendorp, A.V.H., *History of Industry and Trade of the Philippines,* 2 vols., Manila, 1958.

Hendershot, Clarence, "Burma's Value to the Japanese", *Far Eastern Survey,* 11:(1942) 176-8.

Hendershot, "Role of the Shan States in the Japanese

Conquest of Burma", *Far Eastern Quarterly,* II(1943) 253-8.

Hirano, Jiro, "A Study of the Minami Kikan", B.A. thesis, International Christian University, Tokyo, 1963.

Hla Pe, U, *Narrative of the Japanese Occupation of Burma,* Data Paper no. 41, Cornell Southeast Asia Program, Ithaca, 1961.

Huntington, Samuel P., ed., *Changing Patterns of Military Politics,* New York, 1962.

Ike, Nobutaka, *Japan's Decision for War, Records of the 1941 Policy Conferences,* Stanford University Press, 1967.

International Military Tribunal for the Far East, 1946-49, Documents, Exhibits.

Janowitz, Morris, *The Military in the Political Development of New Nations, An Essay in Comparative Analysis,* Chicago, 1964.

Janowitz, Morris, *The New Military, Changing Patterns of Organization,* New York, 1964.

Japan, 16th Army Headquarters, Djakarta, "All Kinds of Armed Bodies", Nishijima papers, Waseda University Social Science Research Institute.

Johnson, John, ed., *The Role of the Military in Underdeveloped Countries,* Princeton, 1962.

Kahin, George McTurnan, *Nationalism and Revolution in Indonesia,* Ithaca, 1970.

Kanahele, George, *The Japanese Occupation of Indonesia: Prelude to Independence,* Cornell Ph.D. dissertation, Ithaca, 1967.

Kelleher, Katherine McArdle, ed., *Political-Military Systems, Comparative Analysis,* Beverly Hills, Sage Publications, 1974.

Kim, Se-Jin, *The Politics of Military Revolution in Korea,* Chapel Hill, 1971.

King, John Kerry, *Southeast Asia in Perspective,* New York, 1956.

Kirby, S. Woodburn, ed., *The War against Japan,* 5 vols., *The Reconquest of Burma,* v.II, London, 1965.

Lebra, Joyce C., "Japan and the Genesis of the Burma Independence Army", *Papers on Far Eastern History,* 5:(1972) 1-35, Canberra, Australian National University.

Lebra, Joyce C., "Japan and Independence Armies of Southeast Asia in World War II", *Bulletin of the Asiatic Society of Japan,* 3 (1972) 1-21.

Lebra, Joyce C., "Japanese and Western Models of the Indian National Army", *The Japan Interpreter,* 7:(1972) 364-76.

Lebra, Joyce C., *Jungle Alliance, Japan and the Indian National Army,* Asia Pacific Press, Singapore, 1971.

Lee, Hahn-been, *Korea: Time, Change and Administration,* Honolulu, 1968.

Lissak, Moshe, "Modernization and Role Expansion of the Military in Developing Countries", *Comparative Studies in Society and History* 9:(1967) 233-56.

McAllister, John T. and Paul Mus, *The Vietnamese and Their Revolution,* New York, 1970.

Mangkupradja, Raden Gatot, "The *Peta* and My Relations with the Japanese: A Correction of Sukarno's Autobiography", *Indonesia,* 5 (1965), Cornell Modern Indonesia Project.

Marr, David G., *Vietnamese Anticolonialism 1885-1925,* Berkeley, 1971.

Maung, Maung, ed., *Aung San of Burma,* The Hague, 1962.

Maung, Maung, *Burma and General Ne Win,* Bombay, 1969.

Maung, Maung, *Burma in the Family of Nations,* Amsterdam, 1956.

Nakamura, Mitsuo, "General Imamura and the Early Period of Japanese Occupation", *Indonesia,* 10 (1970), Cornell Modern Indonesia Project.

Nishijima Shigetada and Kishi Kōichi, eds., *Japanese Military Administration in Indonesia,* U.S. Dept. of Commerce transl., 1965.

Nu, U., *Burma under the Japanese,* London, 1945.

Nugroho, Nototusanto, "The *Peta*-Army in Indonesia", Djakarta, 1971, mimeo.

Nugroho, Nototusanto, "The Revolt of a *Peta* Battalion in Blitar, February 14, 1945", *Asian Studies* 7: (1969) 111-23, Manila.

Ohkawa, J.G., *Two Great Indians in Japan, Sri Rash Behari Bose and Netaji Subhas Chandra Bose,* Calcutta, 1954.

Ohta, Tsunezō, "Japanese Military Occupation of Burma –

the dichotomy", *Intisari,* 2: (*n.d.*) 25-39, Singapore.
Onn, Chin Kee, *Malaya Upside Down,* Singapore, 1946.
Pauker, Guy, "The Role of the Military in Indonesia", Rand Corp., 1960, and in John Johnson, ed., *The Role of the Military in Underdeveloped Countries,* Princeton, 1962.
Penang Shimbun, Georgetown, 1944.
Poppe, James, "Political Developments in the Netherlands East Indies during and immediately after the Japanese Occupation", Ph.D. dissertation, Georgetown University, Washington D.C., 1948.
Prasad, Bisheshwar, ed., *Official History of the Indian Armed Forces in the Second World War, 1939-1945, Reconquest of Burma,* 2 vols., India, 1958.
Price, Willard, *Japan and the Son of Heaven,* New York, 1945.
Price, Willard, *Japan Rides the Tiger,* New York, 1942.
Pye, Lucien, "Armies in the Process of Political Modernization", pp.69-81 in John Johnson, ed., *The Role of the Military in Underdeveloped Countries,* Princeton, 1962.
Quezon, Manuel Luis, *The Good Fight,* New York, 1946.
Ray, J.K., *Transfer of Power in Indonesia,* Bombay, 1967.
Reid, Anthony, *The Contest for North Sumatra; Atjeh, the Netherlands and Britain 1858-1898,* Oxford, 1969.
Robinson, J.B. Perry, *Transformation in Malaya,* London, 1956.
Roeder, O.G., *The Smiling General, President Soeharto of Indonesia,* Djakarta, 1969.
Roff, William, *The Origins of Malay Nationalism,* New Haven, 1967
Sjahrir, Sutan, *Out of Exile,* New York, 1949.
Smail, John R.W., *Bandung in the Early Revolution, 1945-1946; A Study in the Social History of the Indonesian Revolution,* Ithaca, 1964.
Silverstein, Josef, comp., *The Political Legacy of Aung San,* Cornell Southeast Asia Data Paper No.86, 1972.
Silverstein, Josef, ed., *Southeast Asia in World War II:* Four Essays, Yale University Southeast Asia Studies Monograph Series no. 7, 1966.

Singh, General Mohan, *Soldier's Contribution to Independence,* New Delhi, 1974.

Slim Field-Marshal Sir William, *Defeat into Victory,* London, 1958.

Soenarno, Radin, "Malay Nationalism, 1900-1945", *Journal of Southeast Asian History,* vol.1, no.1, 1960: 1-27.

Steinberg, David Joel, *Philippine Collaboration in World War II,* Ann Arbor, 1967.

Sukarno, an Autobiography as told to Cindy Adams, New York, 1965.

Suzuki Keiji, "Aung San and the BIA", in Maung Maung, ed., *Aung San of Burma,* the Hague, 1962.

Thein Pe, M., *What Happened in Burma, The Frank Revelations of a young Burmese Revolutionary Leader who has recently escaped from Burma to India,* Allahabad, 1943.

Thomas, R. Murray, "Educational Remnants of Military Occupation: The Japanese in Indonesia", *Asian Survey* 6:(1966) 630-42.

Thompson, Virginia, "Japan's Blueprint for Indonesia", *Far Eastern Quarterly,* 5: (1945-46) 200-8.

Tinker, Hugh, *The Union of Burma,* London, 1957.

Toye, Hugh, *Subhas Chandra Bose, the Springing Tiger,* Delhi, 1964.

Tregonning, K.G., *A History of Modern Malaya,* New York, 1967.

Tun Pe, U, *Sun over Burma,* Rangoon, 1949.

U.S. Army. *Armed Forces Service Manual.* M354-18A. *Civil Affairs Handbook.* Japan. Section 18A.: Japanese Administration over occupied areas, Burma, 2 Aug. 1944, Research & Analysis Branch, OSS.

U.S. Army. *Armed Forces Service Manual. Civil Affairs Handbook.* Japan. Japanese Administration of Occupied Areas, 18B. Malaya, Research & Analysis Branch, OSS.

Van Niel, Robert, "The Course of Indonesian History", in Ruth McVey, ed., *Indonesia,* New Haven, 1967.

Van Wulften, Palthe, *Psychological Aspects of the Indonesian Problem,* Leiden, 1949.

Van der Mehden, Fred, *Politics of the Developing Nations,* Englewood Cliffs, N.J., 1969.

Van Mook, H.J., *The Netherlands Indies and Japan, Their Relations, 1940-1941,* London, 1944.

Wehl, David, *The Birth of Indonesia,* London, 1948.

Wertheim, W.F., *Indonesian Society in Transition,* The Hague, 1964.

Yoon, Won Z., *Japan's Scheme for the Liberation of Burma: The Role of the Minami Kikan and the 'Thirty Comrades',* Southeast Asia Series no. 27, Ohio University, Athens, Ohio.

Bibliographical Note

This study represents a new area in Western-language scholarship. In nearly all phases of the research I therefore relied on Japanese sources, written and oral.

There are, however, two collections of documents in English which are useful for background material on Japanese occupation policy. They are: Benda, Irikura, and Kishi, *Japanese Military Occupation in Indonesia: Selected Documents;* and Trager, *Burma: Japanese Military Administration, Selected Documents, 1941-1945.*

A number of useful studies and memoirs in English by Southeast Asians deal with the Japanese period. These include Ba Maw, *Breakthrough in Burma;* U Nu, *Japan under the Japanese;* and *Sukarno, an Autobiography as told to Cindy Adams.* U Ba Than's account, *The Roots of Revolution,* was similarly useful. I used in addition memoirs and accounts by members of the INA including Bose, Shah Nawaz, and several civilians.

Published memoirs by Japanese intelligence officers were especially valuable to this study. Those on which I relied most heavily were by Fujiwara Iwaichi, Izumiya Tatsurō, and Yanagawa Munenari. The unpublished account of Sugii Mitsuru on the *Minami Kikan* was also essential.

There are in Japanese two authoritative studies on Japanese military administration in Southeast Asia. These are Ōta Tsunezō, *Biruma ni okeru Nihon gunsei shi no kenkyū,* and the Wasada University-sponsored *Indonesia ni okeru Nihon gunsei shi no kenkyū,* which is also available in a U.S. Department of Commerce translation. I used both versions. The latter, published by Waseda University's Social Science Research Institute, was actually compiled by the late Kishi

Kōichi and Nishijima Shigetada. Nishimima also collected the papers, documents and oral records relating to Japanese military administration in Indonesia which are now at Waseda University's Social Science Research Institute under the auspices of Professor Masuda Atō. Among these are records of interrogation of Japanese officers by returning Dutch authorities after the war. Some of these are in English. There is no parallel collection for any other area of Southeast Asia under Japanese occupation.

Most valuable of all for purposes of this study are the published and unpublished documents and histories in the Boeicho Kenshujo Senshishitsu. Among the published official history volumes prepared by the Senshishitsu which were indispensable are *Biruma kōryaku sakusen* and *Shittan Mei-go sakusen*.

Among the unpublished materials in the Senshishitsu are several catagories of documents essential to this study. One group consists of unpublished diaries, for example excerpts from diaries of Generals Kawabe Shōzō and Inada Masazumi. Another category includes collections of records on Japanese occupation policy toward Southeast Asia. Among these I referred frequently to the *Tokugawa Shiryō* and Ishii *Shiryō,* each of which includes many sets of documents. Beyond this there are a few accounts by Japanese officers of the armies in question, primarily the BIA. The accounts by the late General Sawamoto Rikichirō were indispensable.

Beyond these written records and documents in the Senshishitsu I relied frequently on oral information provided by colonels in the Senshishitsu working on the official history of the War: Colonels Fuwa, Imaoka, and Fukushige.

Since this research is new in English and because records were in some cases unavailable even in Japanese, I of necessity resorted to oral interviews in Japan, India, Thailand, Indonesia, Malaya and Singapore. A list of those interviewed is appended.

I must mention again the Asian scholars who have preceded me in this field and whose work I relied on heavily: Dr. K.K. Ghosh, Mrs. Shiraishi (neé Kurasawa), Colonel Nugroho Nototusanto, and Mr. Minami Jirō.

Interviews

Japan

Lt.-Gen. Fujiwara Iwaichi
Maj.-Gen. Nasu Yoshio
Takahashi Hachirō
Capt. Kawashima Tekenobu
Sugii Mitsuru
Hirano Jirō
Col. Imaoka Yūtaka
Maj.-Gen. Iwakuro Hideo
Hachiya Teruya
Lt.-Gen. Inada Masazumi
Lt.-Gen. Satō Kenryō
Lt.-Gen. Oshima Hiroshi
Lt.-Gen. Katakura Tadasu
Lt.-Gen. Isoda Saburō
Lt.-Gen. Arisue Seizō
Maruyama Shizuo
Kurasawa Aiko
Ishikawa Yoshiaki
Kawadji Susumu
Prof. Ohno Tohrū
Prof. Masuda Atō
Izumiya Tatsurō
Adm. Takagi Sōkichi
Adm. Tomioka Sadatoshi
Gen. Miyamoto Shizuo
Capt. Tsuchiya Kisou
Yoneda Takaichi
Togashi Takeomi
Satō Morio
Adachi Takeshi
Gen. Sakurai Tokutarō
Col. Takeshita Matsuhiko
Horie Yoshitaka
Col. Fuwa Masao
Yamashita Masao
S. Nakaji Seizō
Saitō Munemitsu
Kondo Tsugio

Interviews

India

General Mohan Singh
Col. P.K. Sahgal
Col. G.S. Dhillon
Col. Shaw Nawaz Khan
S.A. Ayer
N. Raghavan
Dr. Girija Mookerjee
Dr. S.K. Bose

Indonesia

Col. Ochiai Shigeyuki
Yanagawa Munenari
Dr. Achmad Subardjo
Col. Zulkifli Lubis
General Bambang Sugeng
Lt.-Gen. P. H. Djatikusmo
General R. Hidajat
Prof. R.H. Kasman Singodimedjo
Col. Nugroho Nototusanto
Effendy Pandjipurnama
Sjachra
Omar Tusin
Dr. Arifin Bey
Brig.-Gen. Subroto Kusmardjo
Soeparjadi

Thailand (Bangkok)

Col. Mya Thaung
Lt.-Gen. L. Hasdintra
Ramlal Sachdev
C.R. Narula
S.T. Mahtani
Dr. N. T. Joseph
Walter L. Meyer
Pandit Raghunath Sharma

Malaysia

Raja Nong Chik
Tunku Abdullah
Raja Shaeran Shah bin Raja Zainil Abidin
Bostam bin Kurshi
Ishak bin Hadji Muhammad
C.C. Too
Eusoffee Abdoolcader

Glossary

Adipadi	Chief of State, dictator, in Burmese
Asrama Merdeka Indonesia	School for a Free Indonesia
Bama Tatmadaw	Burma National Army
Beppan	Abbreviation for *Sambōbu tokubetsuhan,* General Staff Special Section
Biruma Kenkyūkai	Burma Research Association
Bogor *Renseitai*	Bogor Officer Training Unit
bōryaku	stratagem
bundan	squad
chūdan, chūtai	company
Dai Tōa Kyōeiken	Greater East Asia Co-Prosperity Sphere
Dobama Asiayone	We Burmans Society, or Thakin Party
daidan, daitai	battalion
F Kikan	"F" Agency
Gaimushō	Foreign Ministry
gakutai	student corps
giyūgun	volunteer army
giyūtai	volunteer unit
heiho	auxiliary troops
Higashi Hankyū Kyōkai	Eastern Hemisphere Association
Hikari Kikan	"Light" Agency
Hizbullah	Muslim Youth Corps
I-go Kimmutai	First Task Force
Isamu Bunshitsu	*Isamu* Detached Office
Iwakuro Kikan	*Iwakuro* Agency
Java *Hōkōkai*	Java Patriotic Service Association

jibakutai	Suicide Corps
jikeidan	Peace Preservation Corps
kambu kyōikutai	Staff Officer Training Unit
keibōdan	Civil Defense Corps
Keibitai	Police Unit
Kempeitei	Military Police
Kesatwan Melayu Muda	Malay Youth league
Kiai	Muslim religious teacher
Kōain	East Asia Development Board
KRIS	*Kesatuan Raayat Indonesia Semenanjong,* Union of Peninsular Indonesians
laskar rajkat	non-professional soldier, Sumatran name for *giyūgun*
Masjumi Party	*Madjlis Sjuro Muslimin Indonesia*
Miai	organization of *Kiais*
Minami Kikan	Southern Agency
Nami Kikan	Wave Agency
Nampō Kigyō Chōsakai	Southern Enterprises Research Association
Nan'yō Keizai Kenkyūjo	South Seas Economic Research Institute
Nan'yō Kyōkai	South Seas Association
Peta	abbreviation for *Sukarela Tentara Pembela Tanah Air,* Army of Defenders of the Homeland
Poetera, or *Putera*	abbreviation for *Poesat Tenaga Rajkat,* or Center of People's Power
Pusa	*Persatuan Ulama Seluruh Atjeh,* or Central Organization of the Ulamas of Atjeh
Rikugun Nakano Gakkō	Army *Nakano* School, or Army Intelligence School
rimpohan	neighborhood association
rōmusha	forced labor
Sambōbu tokubetsu han	General Staff Special Unit
seinendan	youth group
seishin	spirit

semangat	spirit (Malay)
shodan, shotai	platoon
Sōgō Kenkyukai	General Affairs Research Institute
Shōwa Tsūsho	Showa Trading Company
Suishintai	Pioneer Corps
Sukarela Tentara Pembela Tanah Air	Army of Defenders of the Homeland
Tanggerang Seinen Dōjō	Tanggerang Youth Training Center
Tōa Kenkyūjo	East Asia Research Institute
Tōa Remmei	East Asia Federation
tokumu kikan	special duty agency, or intelligence agency
Thakin Party	Master Party, *Dobama Asiayone*
ulama	Muslim scholars
Yūgekitai	Guerrilla Unit
ulèëbalang	Atjehnese aristocratic class through whom the Dutch ruled Atjeh State

Index

N